Governance in a Changing Market

The Los Angeles Department of Water and Power

Walter Baer

Edmund Edelman

James Ingram III

Sergej Mahnovski

PREPARED FOR THE
**Los Angeles Department
of Water and Power**

ENTERPRISE ANALYSIS

The research described in this report was sponsored by the Los Angeles Department of Water and Power under Contract LADWP04/15/99.

ISBN: 0-8330-2841-3

Published 2001 by RAND
1700 Main Street, P.O. Box 2138, Santa Monica, CA 90407-2138
1333 H St., N.W., Washington, D.C. 20005-4707
RAND URL: http://www.rand.org/
To order RAND documents or to obtain additional information, contact Distribution Services: Telephone: (310) 451-7002; Fax: (310) 451-6915; Internet: order@rand.org

Preface

The Los Angeles Department of Water and Power (DWP), the nation's largest municipally owned electric utility, must prepare for the changes and uncertainties introduced by deregulation, competition, and industry restructuring in the electricity sector. Questions have arisen about whether DWP should continue to operate as a city department or be restructured to compete more effectively in the new environment. Two recent city charter reform commissions considered this issue in 1998 but did not propose significant DWP structural change in the charter amendments presented to Los Angeles voters.

Early in 1999, the DWP asked RAND to conduct an independent analysis of alternative governance structures for DWP as a publicly owned electric utility. Our tasks included reviewing governance changes proposed for DWP; examining how other municipal utilities are structured and governed; and assessing how restructuring would affect DWP, its customers and suppliers, city government officials and agencies, and the city as a whole. We have also tried to place DWP governance in the context of other trends and issues in local government and in the electricity industry.

This report presents the results and findings from our analysis. We hope it will inform discussions of governance in Los Angeles for the many stakeholders in DWP's present and future, as well as present information and options for others concerned with the prospects for municipal utilities in a competitive environment.

Contents

PREFACE iii

TABLES vii

SUMMARY ix

ACKNOWLEDGMENTS xvii

ACRONYMS AND ABBREVIATIONS xix

CHAPTER 1
INTRODUCTION 1
    The Los Angeles Department of Water and Power 1
    Electric Utility Deregulation and Restructuring in California . . . 2
    Choices for Los Angeles and DWP Under Deregulation 3
    Study Purpose and Approach—Outline of This Report 4

CHAPTER 2
THE CURRENT DWP GOVERNANCE STRUCTURE 7
    Board of Water and Power Commissioners 7
    Increased Mayoral Authority over the Commission and DWP . . 8
    City Council Authority and Proposition 5 9
    Legal Representation by City Attorney's Office 10
    State and Federal Authorities Governing DWP 11

CHAPTER 3
DECISIONMAKING AND OPERATIONAL PROBLEMS UNDER THE
    CURRENT STRUCTURE 13
    A Multilayer Reporting Structure for the DWP General
        Manager 14

Hiring and Other Personnel Problems 14
Constraints and Delays in Obtaining Effective Legal
    Representation 15
Cumbersome Procurement and Contracting Procedures 16
Constraints in Negotiating Customer Contracts 19
DWP Financing of Other City Operations 19
Overall Problems of Governance for DWP Competitiveness ... 20

CHAPTER 4
OTHER GOVERNANCE MODELS FOR MUNICIPAL UTILITIES 23
Municipal Utility Reporting to City Council 23
Independent City Agency 24
City-Owned Corporation 25
Municipal Utility District 30
Joint Powers Agency................................. 32

CHAPTER 5
GOVERNANCE OPTIONS FOR DWP 35
Option 1: A City-Owned Corporation to Provide Utility
    Services... 37
Option 2: An Independent City Agency with a Strong
    Governing Board 38
Option 3: Modifications of the Existing Structure to Improve
    DWP Governance................................. 39
Discussion: Rationale for and Objections to Restructuring 42

CHAPTER 6
WHAT COMES NEXT? 45

APPENDIX: A BRIEF HISTORY OF DWP........................ 47

ENDNOTES... 57

BIBLIOGRAPHY... 63

Tables

2.1   DWP Governance Under 1997 and 1999 City Charters . . 12

4.1   Governance Comparisons: DWP and Independent City
      Agencies 26

4.2   Governance Comparisons: DWP, City-Owned
      Corporation, and Municipal Utility District 29

5.1   Governance Options for DWP 36

Summary

The Los Angeles Department of Water and Power (DWP), the largest municipally owned electric utility in the United States, has been the monopoly supplier of electricity to the city's 1.4 million business and residential customers. DWP has provided reliable service, low residential rates, and substantial payments from operating income to the city. However, it now faces major challenges as California proceeds to deregulate, restructure, and introduce competition into the electricity sector.

Since 1998, investor-owned utilities (IOUs), such as Southern California Edison, have been required to offer their customers "direct access" to competitive electricity suppliers. Cities with municipal utilities may decide for themselves whether or not to open their markets, but pressure to allow customers more choices will intensify over the next several years. In response, DWP has implemented a series of measures to reduce its operating costs and has set a goal of paying off all or most of its debt on generating plants by 2003. At that point, DWP would be better prepared to compete with other electricity suppliers if the city council decides to open the Los Angeles market.

However, DWP's general manager and others have questioned whether DWP, organized as a city department and subject to the checks and balances of city governance, will be able to compete effectively. This report examines DWP governance issues in the context of electricity deregulation and restructuring and discusses alternative structures for governing DWP as a municipally owned utility. The study explicitly does not consider privatization or sale of DWP electric power operations or assets.

THE CURRENT DWP GOVERNANCE STRUCTURE

Governance of the DWP is shared among the Board of Water and Power Commissioners, the office of the mayor, the city council and its staff, and the city attorney. In effect, the DWP general manager must report to all these entities, which may themselves have conflicting objectives. Seventy-five years ago, the city charter established a strong commission with primary authority to oversee the department. But through charter amendments passed over the past two decades, the mayor and council have gained more control at the expense of the commission.

The mayor has the power to appoint and remove Water and Power Commissioners. The council must confirm each appointment and removal by majority vote, but under new charter amendments that went into effect on July 1, 2000, the mayor may remove a commissioner without council approval. It has become customary for newly elected mayors to appoint their own commissioners and remove unwanted holdovers. While justified politically as the way for the city's top elected official to establish control over the DWP and other city departments, this effectively has vitiated the commission as an independent, nonpolitical governing board. The mayor also holds tight rein over the commission through "advice" from his staff, and by requiring approval of commission agenda items by the city administrative officer under Executive Directive 39.

As the city's legislative body, the council has both oversight responsibility for DWP and direct authority under the charter to approve rates, set job classifications and compensation under the city's civil service system, approve property sales, and approve contracts of more than $150,000 or of more than three years in duration. The council has traditionally set electricity rates to benefit residential customers—i.e., voters. Council ordinances further specify, in considerably more detail, procedures for hiring and other personnel actions, issuing debt, contracting, negotiating long-term customer contracts, and many other operational matters. But the most controversial of the council's authorities over DWP comes from a charter amendment known as Proposition 5, or "Prop. 5," which allows the council to reconsider essentially any decision made by the commission. The threat of Prop. 5 veto further undermines the commission's ability to exercise independent judgment in overseeing the DWP and results in more bureaucratic paperwork and delays in decisionmaking.

The elected city attorney serves as legal advisor to the commission and DWP. The city attorney's office provides the department's legal staff and is responsible for making personnel and work assignments. Lawyers working on DWP legal matters report to the city attorney rather than to the DWP general manager or the commission. Moreover, the city attorney must approve any use of outside counsel.

Decisionmaking and Operational Problems Under the Current Structure

Not surprisingly, this divided governance structure complicates and slows down commission and DWP decisionmaking, as well as the department's ability to take timely action. Specific issues include

- a multilayer—often conflicting—reporting structure for the general manager;
- constraints and delays in hiring managers, professionals, and skilled workers;
- constraints and delays in obtaining effective legal representation;
- cumbersome procurement and contracting procedures;
- constraints in negotiating customer contracts; and
- politically driven DWP financing of other city operations.

The current structure was put in place deliberately to provide extensive checks and balances for a government department that had a monopoly on providing essential water and power services. Delays in making or implementing business decisions were of less concern and have not had much adverse impact on DWP revenues or profitability during the monopoly era. But they can make the difference between winning and losing customers in a competitive market. DWP business customers who now subsidize residential rates are particularly apt to respond to lower cost or more flexible alternatives. A substantial loss of business customers would inevitably lead to reduced DWP operating income and strong pressure on the mayor and council to cut either DWP payments to the city or raise residential electric rates, or both.

OTHER POSSIBLE GOVERNANCE MODELS FOR DWP

Many municipal utilities in North America have different and generally more streamlined governance arrangements. Such models include the following:

- Direct reporting to the city council (e.g., Austin, Texas; Colorado Springs, Colorado).
- Independent city agency (e.g., Jacksonville, Florida; Knoxville, Tennessee).
- City-owned corporation (e.g., Toronto, Ontario; Safford, Arizona).
- Municipal Utility District (e.g., Sacramento, California).
- Joint Powers Agency (e.g., Southern California Public Power Authority).

Direct reporting to the city council works well in smaller cities but does not seem appropriate for a utility as large and complex as the DWP or for a city as diverse and fractious as Los Angeles. A Municipal Utility District or a Joint Powers Agency would offer flexibility and independence in conducting day-to-day operations; but each would require new state legislation as well as local restructuring and thus might be difficult to achieve politically. If passed by county voters, however, a Municipal Utility District would be more difficult to undo than the other options.

For DWP, operating as a city-owned corporation or as a more independent city agency appear promising alternatives to the status quo. Either one would help the utility become more efficient, businesslike, and responsive to changing market conditions. However, both involve substantial restructuring and new charter amendments to invest primary governance responsibility in a single board. With corporatization, an expanded and independent board of directors would govern the utility. In the city agency option, a similarly expanded and more independent commission would serve as the governing board. Restructuring would give the new board considerably more authority to oversee the utility and would deliberately distance the board and utility from oversight by the mayor and council on normal business matters.

A third, but probably less effective, option is to modify the existing governance structure to improve DWP decisionmaking and operations. The goal would be to focus governance on policy issues, limit involvement in routine operations, and streamline approval processes. Proposed changes in this option would include

- authorizing water and power commissioners to serve full five-year terms;
- enabling the commission and DWP to hire its own legal advisor and staff;
- eliminating formal executive review of commission agenda items; and
- eliminating council oversight of DWP routine business and giving DWP more flexibility in procurement, contracting, and personnel matters.

The first two changes would require new charter amendments, but the mayor and council could implement the latter two within the existing structure.

All three options would maintain the primary public benefits of a municipal utility: local ownership and local rate setting authority; tax exempt financing and preferences in purchasing federal power; exemption from most income, property, and business taxes; sensitivity to economic development, environmental concerns, and other social goals; and commitment to make direct transfers to the city's general fund. Each option also keeps the utility's governing board accountable to elected city officials.

DEREGULATION AND FUTURE GOVERNANCE FOR DWP: FINDINGS AND RECOMMENDATIONS

Electricity deregulation has recently come under fire as higher demand for power in California brought blackouts in the Bay Area, sharp spikes in the wholesale price of electricity, and a doubling of retail prices for customers of San Diego Gas & Electric, the first investor-owned utility to be fully deregulated. Retail prices for customers of other IOUs also may rise as the transitional rate ceilings under the state deregulation plan phase out by the end of 2001. As a consequence, customers and government officials fought successfully this summer to cap wholesale spot prices, and some are demanding that deregulation be rescinded or at least substantially revised.

Throughout the crisis, DWP not only kept rates stable and continued to reduce debt, but it has earned substantial profits by selling electricity it generates to the California Independent System Operator (ISO). Whereas the IOUs sold off their generating facilities under the state deregulation plan, DWP still maintains substantial reserves. A recent opinion piece in the *Los Angeles Times* called DWP "The Unexpected Hero in a Deregulated Electricity Market" and recommended that the city "say 'no' to deregulation."

Is it necessary or desirable to deregulate electricity prices and permit direct access competition in Los Angeles? In the short run, the answer is clearly no. But in the longer run, if and when the wild price fluctuations observed this summer settle down and orderly electricity markets again become the norm, the question will arise again. We believe this is likely to occur in the 2002–2004 time frame as new generation capacity comes into service in California, and the IOUs and the California ISO gain more experience in stabilizing electricity markets. While city, state, and federal governments stand ready to act to protect consumers from large, short-term price increases, we do not expect them to reverse the underlying trend toward encouraging more competition among electricity suppliers.

Whether or not the Los Angeles electricity market is opened to competition, DWP's governance needs simplification and streamlining. We believe that the restructuring options of corporatization or governance by a strong commission deserve serious attention. Under its current governance structure—even with the modifications recommended in the third option described above—DWP would find itself severely constrained in meeting the competition from more agile private firms that we expect to emerge around 2002–2003. In principle, a city-owned corporation could have more operational flexibility than an independent city agency, but either form would support faster decisionmaking and greater responsiveness than does the present structure. Establishing a single governing board, with clear authority and considerable independence from day-to-day political influences, seems a prerequisite for success in a more competitive marketplace.

The issues of DWP governance and possible restructuring are necessarily linked to the council's consideration of whether or when to open the Los Angeles electricity market. Even if the council's decision on direct access is not made until 2002 or later, discussion and debate should begin soon. The public needs to become aware of issues that, while often technical and complex, will directly affect them as taxpayers and ratepayers. The effects of competition on DWP, its employees, its customers, and the city as a whole need to be more fully explored. Possible charter amendments need to be vetted by the council before they can be put before the voters. These issues also are likely to arise in city council races and in the mayoral election of 2001.

Whatever the council's decision on direct access competition, DWP must improve its decisionmaking pace and processes. It must run faster in the future to stay competitive. Strengthening its governance structure seems essential to ensuring reliable electricity supplies, low rates, and adequate payments to the city, as well as to maintaining Los Angeles's leadership among the nation's municipal utilities.

Acknowledgments

We are grateful to many people who assisted us throughout this study. S. David Freeman, Frank Salas, Steve Sugerman, and Eric Tharp opened many doors for us with introductions to those, both inside and outside DWP, who were knowledgeable about the department, Los Angeles city government, and the issues surrounding DWP governance. They also responded without complaint to our incessant requests for more information.

We particularly want to thank the individuals we interviewed during the course of our research, including public officials, utility executives, union leaders, and others who graciously shared their time and knowledge with us. This document also benefited greatly from a thorough and constructive review of an earlier draft by Professor Steven P. Erie of the University of California, San Diego.

Acronyms and Abbreviations

AB	Assembly Bill
CalPX	California Power Exchange
CAO	City administrative officer
CEO	Chief executive officer
CLA	Chief legislative analyst
CTC	Competition transition charge
DWP	(Los Angeles) Department of Water and Power
FERC	Federal Energy Regulatory Commission
GAAP	Generally accepted accounting principles
GO	General obligation
IOU	Investor-owned utility
ISO	Independent System Operator
JPA	Joint Powers Agency
LAFCO	Local Agency Formation Commission
MUD	Municipal Utility District
NCPA	Northern California Power Authority
SCPPA	Southern California Public Power Authority
SMUD	Sacramento Municipal Utility District

Chapter 1
Introduction

The Los Angeles Department of Water and Power

The Los Angeles Department of Water and Power (DWP) is the largest municipally owned electric utility in the United States. In 1999 its electricity operations served 1.4 million customers, sold 27 billion kilowatt hours, generated revenues of $2.3 billion, and transferred $129 million to the City of Los Angeles (DWP, 1999).[1] Unlike many other municipal utilities that solely distribute power purchased from others, DWP has substantial ownership shares in hydroelectric, fossil fuel, and nuclear generating plants, as well as in the high-voltage transmission lines serving Southern California.

The City of Los Angeles began generating hydroelectric power in the early 1900s as a by-product of the Owens Valley Aqueduct project built by the city Water Department.[2] In 1911, voters approved establishing a municipal Power Bureau to distribute electricity within the city rather than selling it to private power companies, such as Southern California Edison (SCE). City charter amendments in 1925 created the Water and Power Department headed by a five-member citizen commission. This basic governance structure continues today, although (as discussed in the following section) the commission's and department's relationships with the mayor, council, and other city officials have become considerably more complex.

As a municipally owned utility, DWP can issue tax-exempt revenue bonds and has preference over investor-owned utilities in purchasing low-cost power from federal generating projects. It does not pay federal or state income taxes or Los Angeles property and business taxes but instead makes direct payments from operating income to the city's general fund. The city council has traditionally set utility rates to benefit Los Angeles residential customers—i.e., voters. DWP's residential electricity rates today are about 11 percent lower than those of SCE, its for-profit neighbor, while its industrial and commercial rates are about 15 percent higher (DWP, 1999; Edison, 1999).

DWP's reputation as a reliable and efficient electricity supplier has played an important role in the growth of Los Angeles for most of this century. However, it now faces major challenges as competition in the electricity sector comes to California.

ELECTRIC UTILITY DEREGULATION AND RESTRUCTURING IN CALIFORNIA

Electricity is following a path similar to the experiences of the telephone and gas industries in the 1980s and 1990s: deregulation, competition, and restructuring.

Electric utilities in the United States have historically been vertically integrated, regulated monopolies that control all aspects of electricity supply and distribution. Under pressure from regulators and customers for lower costs and greater choice, however, the monopoly structure is crumbling.

In electricity, restructuring takes the form of separating the three components of electricity supply: generation, transmission, and local distribution. Generation becomes a fully competitive market. With deregulation of the generation component, customers can buy electricity—whether from a faraway coal power plant or a nearby gas turbine—from suppliers other than their local utility. This is usually referred to as "direct access" competition. Transmission from the power source to the local distributor becomes a separate industry sector regulated by an Independent System Operator (ISO). Distribution to the customer generally remains a monopoly service provided by a local utility regulated primarily by the state Public Utilities Commission.

California leads most other states in deregulating and restructuring the investor-owned utility (IOU) sector that serves about 70 percent of the state's electricity customers. Restructuring is mandated by Assembly Bill (AB) 1890, which became law in September 1996. SCE and other California IOUs began offering competitive direct access to their customers on April 1, 1998. AB 1890 created a California Power Exchange (CalPX), providing a spot market for electricity sales and purchases, and it required IOUs to make their high-voltage transmission facilities available on a fair and equitable basis under the supervision of the new California ISO. With electricity prices determined by the market, some generating plants can no longer compete economically and must be taken out of service before they are fully depreciated. AB 1890 permitted IOUs to recover such "stranded investments" through special "competition transition charges" levied on customers through March 31, 2002.

CHOICES FOR LOS ANGELES AND DWP UNDER DEREGULATION

AB 1890 mandated restructuring and direct access competition only for the IOUs. The municipally owned utilities, which serve 30 percent of the state's customers, are not required to offer direct access competition or to participate in the CalPX and ISO. However, each city council or other utility-governing body must hold public hearings and then make a formal decision on whether or not to open its market and give utility customers direct access to competitive suppliers.

In November 1997, at the request of the mayor and city council, DWP's new general manager, S. David Freeman, prepared an "Action Plan to Meet the Competitive Challenge." It included proposals to freeze residential rates, downsize the workforce, pay down debt, and take other measures to reduce operating costs (DWP, 1997). These programs have largely been implemented. DWP has reorganized into business units for generation, transmission, and distribution. It joined the CalPX in December 1998. By 2003, the department expects to have paid down all or most of its debt on generating plants that might represent "stranded costs" under competition.[3] At that point, DWP would be better prepared to compete with SCE and other electricity suppliers if the council decided to open the Los Angeles market.[4]

But is DWP, organized as a city department and subject to all the checks and balances of city governance, lean and agile enough to succeed in a competitive market? Does competition require restructuring the present governance system set forth in the city charter? In his remarks to the Los Angeles Charter Reform Commission in March 1998, General Manager David Freeman argued that restructuring is necessary to increase the tempo of decisionmaking at DWP and cut bureaucratic delay (Charter Reform Commission, 1998b). As he elaborated in a subsequent interview: "The current structure is designed to control a monopoly. . . . if you deregulate my competitors and allow them to make deals with my customers on a daily basis . . . I can't compete as long as the monopoly controls continue to exist." (Metro, 1998.)

Others, in and out of city government, have been less convinced. The charter reform measures approved by Los Angeles voters in June 1999 made only minor adjustments to the DWP governance structure. And the Los Angeles City Council has not yet formally addressed the question of opening the city to direct access electricity competition.

STUDY PURPOSE AND APPROACH—OUTLINE OF THIS REPORT

At DWP's request, this study was undertaken in April 1999 to examine DWP governance issues in the context of electricity deregulation and restructuring and alternative structures for governing DWP as a municipally owned utility. The study explicitly did not consider the privatization or sale of DWP electric power operations or of its generating and transmission components.[5] The analysis included

- a literature review on electricity deregulation, restructuring, and competition in the United States, United Kingdom, and other countries;
- a review of AB 1890 and other state and federal legislation and regulations affecting electricity deregulation, restructuring, and competition in California;
- a review of DWP governance issues raised by the Los Angeles Charter Reform Commissions (Charter Reform Commission, 1998a; Elected Charter Reform Commission, 1998a, 1998b, 1998c, and 1998d);[6] and in other previous studies (Metzler, 1990; Barrington-Wellesley, 1994; Beck, 1996a; and PSC, 1996);

- interviews with current and former DWP managers; current and former Water and Power Commissioners; city council members and staff; mayor's office, city administrative officer (CAO), chief legislative analyst (CLA), and city department staff; labor and business leaders; and academics, attorneys, consultants, journalists, and other knowledgeable Los Angeles stakeholders; and
- interviews with executives of other municipal and investor-owned utilities in the United States and Canada.

Chapter 2 describes the current DWP governance structure and the changes that will take effect next year under the new city charter amendments. Chapter 3 discusses DWP decisionmaking and operational problems under the current structure and suggests how competition may exacerbate or otherwise affect them. Chapter 4 then outlines other municipal utility structures in the United States and Canada and compares them with DWP. Three preferred options for modifying or restructuring DWP governance are presented in Chapter 5, followed by a short final chapter outlining some next steps toward deciding how DWP will be structured and governed in the twenty-first century.

Chapter 2
The Current DWP Governance Structure

Governance of the Department of Water and Power is shared among the Board of Water and Power Commissioners, the office of the mayor, the city council and staff, and the city attorney. In effect, the DWP general manager reports to all of them, albeit in different ways on different policy and operational issues. This chapter describes the complex interactions among these governing entities, as well as the roles of such other important actors as the city controller, the CAO, and the CLA. It also outlines how the new city charter amendments, which were adopted in June 1999 and went into effect in July 2000, will affect DWP governance.

BOARD OF WATER AND POWER COMMISSIONERS

The 1925 Los Angeles City Charter established a five-member Board of Water and Power Commissioners to head the DWP. Commissioners were appointed by the mayor and confirmed by the city council to five-year, staggered terms. The commission selected its own officers from among its members, chose the general manager,[7] and generally was empowered to oversee the department.[8]

Because the DWP generates its own revenue from water and power sales, the 1925 charter established it as a "proprietary department" with somewhat more autonomy than other city departments. The DWP has its own budget that is separate from the city's general fund, can hold property separate from the city, and can issue debt backed by its own revenue rather than rely on the city's general obligation bonds. For more than 50 years, the commission could set salaries for DWP employees covered under the city's civil service system, but this authority passed to the council in 1977.

INCREASED MAYORAL AUTHORITY OVER THE COMMISSION AND DWP

Although the Water and Power Commission initially had strong executive power over the department, a series of changes since the 1960s have reduced its authority and placed it and the DWP under the control of the city's elected officials.

The mayor exerts principal authority by appointing and removing Water and Power Commissioners. The council must confirm each appointment and removal by majority vote, but under the new charter amendments, the mayor may remove a commissioner without council approval (New Charter, 1999, Section 502(d)).[9] Equally important, it has become customary for newly elected mayors to appoint their own commissioners and remove unwanted holdovers. While justified politically as the way for the city's top elected official to establish control over the DWP and other city departments, this effectively has vitiated the commission as an independent, nonpolitical governing board. Moreover, the mayor currently selects the DWP general manager, with council confirmation. The new charter returns CEO selection to the commission, but subject to approval by both mayor and council.[10]

The mayor holds tight rein over the commission through staff "advice" to commissioners on specific agenda items. The mayor also requires approval of commission agenda items by the CAO under Executive Directive 39 (ED39). Originally issued by Mayor Tom Bradley in 1984, ED39 has been used by Mayor Richard Riordan to control the commission agendas of DWP and the other proprietary departments. Although ED39 officially requires review only of important policy-relevant items—including proposed charter amendments and ordinances, contracts and leases with policy implications, bond and debt orders, changes in rates or fees, and major organizational changes—it in fact is applied to other commission matters.[11] The two principal arguments for CAO review of commission agendas are to coordinate proposals from different departments and to give a citywide perspective on matters that must ultimately go to the city council.[12] However, ED39 review also has the effect of slowing commission decisionmaking and adding another layer of bureaucracy to DWP's already-cumbersome approval processes.[13]

CITY COUNCIL AUTHORITY AND PROPOSITION 5

As the city's legislative body, the city council has both oversight responsibility for DWP and direct authority under the city charter to approve certain commission actions, including

- tariffs, rates, and other charges;
- job classification, compensation, and other aspects of the civil service system;
- real property sales and leases of more than five years;
- contracts of more than a fixed amount (currently $150,000) or lasting longer than three years;
- initial authority to issue debt;
- participation with other public or private parties in major capital projects; and
- proposed ordinances or charter amendments affecting DWP.

Council ordinances further specify, in considerably more detail, procedures for hiring and other personnel actions, issuing debt, contracting, negotiating long-term customer contracts, and many other day-to-day operational matters. The chief legislative analyst acts as the council's agent on many issues and often has de facto authority in dealing with DWP.[14]

But the most controversial of the council's authorities over DWP comes from charter Section 32.3, generally known as "Prop. 5" after the measure's designation on the June 1991 ballot. Prop. 5 provides for five council meeting days in which any action[15] by the Board of Water and Power Commissioners can be taken up for reconsideration by the council by a two-thirds vote. Prop. 5 gives the council three weeks to substitute its decision, by simple majority vote, for that made by the commission. If the council does not make its decision during this period, the commission's action becomes final.

Prop. 5 basically permits the council to intervene in any major or minor policy or operational aspect of the department's business and affairs. According to a May 1998 study by the mayor's office, 34 matters involving the Board of Water and Power Commissioners had been taken up by the council under Prop. 5 (Riordan, 1998). This represented nearly one-third of the total number of "Prop. 5-ed" items. The council substituted its judgment for the board's in only 4 of the 34 matters; three commission decisions were overturned, and a fourth was moved to the council's direct jurisdiction.[16]

Although Prop. 5 has not been used often to overturn Water and Power Commission decisions, it has had a demoralizing effect on both the commission and the department. The threat of Prop. 5 has further undermined the commission's ability to exercise independent judgment in overseeing the DWP. Moreover, the need to buttress even minor matters against the threat of Prop. 5 repudiation appears to have led to increased paperwork and substantial delays in decisionmaking.

As a result of the new charter amendments that went into effect in July 2000, the council's ability to substitute its own action for that of a commission has been replaced by a legislative veto (New Charter, 1999, Section 245). That is, the council is able only to remand the action back to the board for reconsideration and a new action. These changes may reduce the temptation for the council to challenge commission decisions, although how they will work in practice remains to be seen.[17]

LEGAL REPRESENTATION BY CITY ATTORNEY'S OFFICE

The Board of Water and Power Commissioners and DWP do not hire their own legal staff. The city charter gives the city attorney the role of representing the board in litigation and acting as the board's legal advisor (Old Charter, 1997, Section 42; New Charter 1999, Section 271). The city attorney's office provides legal staff to the department and makes work assignments. Lawyers working on DWP matters report to the city attorney rather than to the DWP general manager or the commission. Upon recommendation by the commission, and with the written consent of the city attorney, "the city may contract with attorneys outside of the city attorney's office to assist the city attorney in providing legal services" to DWP (New Charter, 1999, Section 275).

This arrangement has led to conflicts about who is the real client on DWP legal matters: the commission and department or the city as a whole? City attorneys, who are elected citywide by the voters, have typically taken the position that they and their staff represent the city at large. But commissions and DWP general managers contend that the city attorney must represent them as the clients on DWP legal matters. The new charter comes down on DWP's side, stating that the "boards of the Proprietary Departments . . . shall make client decisions in litigation . . . [and] shall have the authority to approve or reject settlement of litigation exclusively involving the policies and funds over which the charter gives those boards control." (New Charter, 1999, Sections. 272 and 273.) But on a day-to-day basis, the city staff attorneys do not take direction from DWP.

Table 2.1 provides a summary of the multilayer municipal governance structure for DWP and the changes introduced by the new charter amendments.

STATE AND FEDERAL AUTHORITIES GOVERNING DWP

In addition to governance at the municipal level, the Water and Power Commission and the DWP are subject to a variety of federal and California laws and regulations. Among the most important of these are the following:

- California's Brown Act, which requires such public bodies as the commission to hold all its meetings openly in public.
- California's Meyers-Milias-Brown Act, requiring the city to meet in good faith with union leaders for collective bargaining purposes.
- The provisions of California AB 1890 regarding electricity restructuring and competition as they pertain to municipal utilities.
- Regulations of the Federal Energy Regulatory Commission (FERC) regarding access and interconnection to high-voltage transmission lines.
- Federal tax statutes and regulations that place limitations on the use of municipal utility facilities financed by tax-exempt debt.

Table 2.1
DWP Governance Under 1997 and 1999 City Charters

Governance Structure	1997 City Charter	1999 Charter Changes
Organization	City department	
Governing board Term of office	Five-member commission; staggered, five-year terms	
Selection	Mayor appoints, council majority confirms	
Removal	Mayor may remove with council majority approval	Mayor may remove without council approval
Authorities retained by city council (partial list)	Approval of rates	
	Job classification, compensation	
	Approval of contracts, sales, leases	
	Authorization of new debt	
	Approval of joint projects	
Other limits on board authority	Council may reconsider and change any commission decision (Prop. 5)	Council may veto but not change commission decisions (Prop. 5)
	Open meetings required (Brown Act)	
	CAO must approve agenda items (ED39)	
General manager selection and removal	Mayor appoints, council majority confirms	Board appoints with mayor and council approval
	Mayor may remove with council majority approval	Board may remove with mayor's approval, but two-thirds council vote can reinstate
Other employee status	Civil service except for up to 16 exempt positions that require mayor and council approval	Up to 16 DWP exempt positions approved by mayor unless overruled by two-thirds council vote within 10 days
		Up to 150 additional exempt positions authorized for city as a whole, including DWP
Legal staffing and litigation decisions	City attorney provides legal staff	Board makes client decisions in litigation and settlement
	Use of outside legal counsel requires written approval of city attorney and council	Use of outside legal counsel requires written approval of city attorney
	Council controls litigation with city attorney acting on city's behalf	

Chapter 3
Decisionmaking and Operational Problems Under the Current Structure

As indicated in the previous chapter, DWP's governance structure is complex, divided, and cumbersome. Many operational and management decisions must be reviewed sequentially by different city bodies, including the commission, the CAO and other mayoral staff, the city attorney's office, and the city council and CLA. This structure was put in place deliberately to provide extensive checks and balances for a government department that had a monopoly on providing essential water and power services. Delays in making or implementing business decisions were of less concern and have had little adverse impact on DWP revenues or profitability during the monopoly era. That is changing as electricity restructuring moves forward in California and pressure builds to open the Los Angeles market to direct-access competition.

This chapter discusses some of the problems evident under the current governance structure that could seriously hamper DWP in a more competitive environment. They include

- a multilayer reporting structure for the general manager;
- constraints and delays in hiring managers, professionals, and skilled workers;
- constraints and delays in obtaining effective legal representation;
- cumbersome procurement and contracting procedures;
- constraints in negotiating customer contracts; and
- DWP financing of other city operations.

A Multilayer Reporting Structure for the DWP General Manager

While the charter states that the DWP general manager works under the "instruction of his or her board" of Water and Power Commissioners (Old Charter, 1997, Section 80(a)), he or she also reports to the mayor and council. In fact, the commission appears to have the weakest reporting relationship on such important matters as staffing, negotiating with customers, and resolving legal disputes. This complex and divided reporting structure severely limits the general manager and top-level DWP staff in their ability to make and implement operational decisions in a timely way. In sharp contrast, CEOs of IOUs, as well as some successful municipal utilities (described in Chapter 4), report to a single strong governing board.

Recruiting and retaining an experienced general manager is difficult today for any municipal utility because of the much larger salaries, stock options, and other incentives offered by private firms. Recruiting at DWP faces the additional burden of the multilayer reporting arrangement for the general manager. As DWP's current general manager, S. David Freeman, puts it: "Everybody's in charge and nobody's in charge. I don't have a Board of Directors . . . I have two Boards and a Mayor. And sometimes there're differences of opinions among them." (*Metro*, 1998; Freeman, 1998.)

Hiring and Other Personnel Problems

Hiring at DWP is complicated by the fact that nearly all the department's more than 7,200 employees fall under the city's civil service system. This means prospective employees generally must pass a civil service exam before they can be hired—a procedure that can add weeks or months to the hiring process. Such delays are especially painful when trying to hire technically skilled workers, who are in great demand from other employers and often will not wait to qualify under the city's civil service rules. In our interviews, we heard that these problems had arisen in DWP's efforts to hire skilled workers for the Valley Steam Plant, as well as staff for information technology and marketing positions. In contrast, we were told, other utilities that are subject to union agreements but not civil service rules can hire people with similar skills within a few weeks.

Management and professional staff hiring at DWP is also hampered by the relatively low salaries and rigid job categories that the civil service system applies to all city departments. DWP is highly constrained in its ability to promote or offer financial incentives to superior performers or to demote or dismiss poor performers. In a dynamic economy, such civil service constraints outweigh the benefits of job protection for those with salable skills.

Only 15 DWP positions (in addition to the general manager and the chief financial officer) are exempt from civil service rules.[18] Filling these positions is today a two-step process. First, the council must pass a resolution by two-thirds vote, "which sets forth the educational experience and other professional requirements of the position(s)," as well as "the circumstances . . . that preclude filling the position(s) through the civil service system." (Old Charter, 1997, Section 111.) Once a candidate has been selected, he or she must be formally approved by both the mayor and council. This process can take several months or longer. As a consequence, DWP often finds itself unable to compete effectively in hiring top-flight managers and professionals with experience in the electric utility business.

The new charter simplifies the approval process by replacing the council resolution with a recommendation by the mayor proposing the qualifications for the exempt position. If the council does not veto the recommendation by two-thirds vote within 10 council meeting days, it is deemed approved (New Charter, 1999, Section 1001).[19] While this should shorten the time involved, it does not change the basic requirement for both mayor and council approval of every DWP exempt position.

Constraints and Delays in Obtaining Effective Legal Representation

As described in Chapter 2, DWP's legal matters are handled by the city attorney's office. The department cannot hire or choose its own legal staff, except in special circumstances approved in writing by the city attorney and the council.

While the city lawyers assigned to DWP can and do handle routine department matters effectively, they often do not have the expertise to deal with the growing number of complex energy and environmental issues. According to those we interviewed, obtaining the necessary approvals to hire outside counsel takes several weeks at best, which can seriously impede the department in negotiations or other time-sensitive legal matters. Moreover, the city lawyers assigned to DWP report to, and owe their allegiance to, the city attorney. DWP management cannot replace or reassign them if it does not like their work. City attorney staff are thus not as responsive to DWP priorities and urgency (e.g., working at night to finish a contract or filing) as attorneys working directly for the department or in private practice would be. And in some cases, the views of the city attorney, an elected official who represents the city as a whole, may be at variance with DWP's views.

Two legal issues that illustrate these problems, cited by DWP General Manager David Freeman, involve disputes with Owens Valley over air quality and with Montana Power over an energy supply contract. After becoming general manager late in 1997, Mr. Freeman sought to settle the long-standing Owens Valley conflict, contrary to the position then held by the city attorney staff representing the department. A settlement was reached in July 1998 after DWP received permission to hire outside counsel with expertise in air quality issues. The Montana Power dispute also required specific expertise, which the city attorney staff lacked. According to department sources, Montana Power's lawyers "did not take us seriously" until DWP brought in outside counsel. In both cases, obtaining permission took several months and required the personal intervention of the general manager.

Cumbersome Procurement and Contracting Procedures

Like all city departments, DWP must abide by complex procurement and contracting practices designed with many checks and balances to minimize abuses and serve other public objectives. For good political reasons, procurement is generally slower, less flexible, and more expensive in the public than in the private sector. However, DWP purchases more goods and services than other departments do, so its burden arising from operating under city procurement regulations is commensurately greater. While this burden has not been critical in a monopoly environment,[20] it will hinder DWP's ability to compete with more-agile for-profit firms. Primary concerns include the cost and delay built into current contracting procedures, and the demands on suppliers over and above those required by other utilities.

A standard DWP procurement for less than $150,000 goes through the following steps:

- Requisition by a DWP unit.
- Preparation of a bidding document.
- Formal request for bids sent out (and advertised if the requisition is for more than $25,000).
- Responses to questions from potential bidders.
- Opening, recording, and posting of bids.
- Technical evaluation by requesting unit.
- Nontechnical evaluation by Procurement Department.
- Recommendation for award by Procurement Department.
- Award decision by DWP management if under $150,000.
- Notification of the successful bidder.
- Contract drafting and signing.

Procurement in the for-profit world goes through similar steps, but the process is much less formal and more flexible than at DWP. Firms, for example, usually negotiate with prospective suppliers for better prices and terms, but under city contracting procedures, DWP cannot negotiate and must award the contract to the "lowest and best regular responsible bidder." (DWP, 1999a.)

For amounts above $150,000, the process becomes even more complex. All DWP purchases above $150,000 must be formally approved by the Water and Power Commission.[21] In effect this means they must also be approved by the mayor's staff (in order to go on the commission agenda, per ED39) and by the council (through the Prop. 5 process).[22] Between 200 and 300 DWP procurements each year go through this extended review process, and the commission must consider an average of about a dozen contract awards at each of its meetings.

This multilayered approval process costs both the department and the commission considerable time and money. The requesting DWP unit must write a formal letter recommending and justifying the procurement. Increased time is allotted for advertising, bidding, and evaluation. The DWP Procurement Department must then prepare a well-documented contract-justification package for approval by the commission (and possible review by the mayor and council). The item must be placed on the agenda of a regular commission meeting, which occurs every two weeks. After commission approval, DWP must wait five council meeting days to see whether the council will take up the item under Prop. 5. Overall, assuming no further Prop. 5 review, DWP staff estimates that it takes 90–120 days to award a contract of more than $150,000, about twice the time it takes to award a smaller one. The actual dollar costs are difficult to document, but, based on DWP staff accounts of the added time to prepare a commission procurement package and manage the process, we estimate that the additional cost to DWP runs $2 million to $3 million a year.

Another constraint arises from an ordinance requiring that once the DWP annual budget is passed by the council, any transfers among internal budget accounts greater than $35,000 (or 1 percent of the account, up to a maximum of $100,000) must be individually approved in writing by the mayor. The city also places a number of compliance requirements on DWP contractors. Before receiving an award, DWP contractors must obtain a City of Los Angeles Business Tax Registration Certificate and certify compliance with (among others) affirmative action and equal opportunity programs, the Los Angeles Child Support Obligations Ordinance, the Los Angeles Living Wage Ordinance, and the Los Angeles Service Contract Worker Retention Ordinance. Suppliers to corporations, including potential competitors of DWP, generally do not have to comply with these city requirements.

Constraints in Negotiating Customer Contracts

DWP's industrial and commercial customers are the most likely to respond to competitive offerings, because their average rates now are higher than those offered by SCE in surrounding areas.[23] However, customer surveys repeatedly find that power reliability is even more important than price for most businesses and that DWP retains its superior reputation for reliability.[24] DWP still must remain price competitive and be able to respond quickly to competitors' efforts to woo customers away.

Until recently, the department's ability to negotiate with customers was constrained by the need for council approval of every such arrangement. Consequently, at DWP's request, the council passed an ordinance giving DWP authority to offer discounts of up to 5 percent to customers who sign long-term contracts for up to 10 years. DWP recently announced that 22 large customers—including the Los Angeles Unified School District, McDonald's Corporation, Robinsons May, Kaiser Foundation Health Plan, Anheuser-Busch, and the Getty Center—have signed such long-term contracts (DWP, 1999c).

Many at DWP, and especially the general manager, believe that when competition comes to Los Angeles, the department will need considerably more flexibility to negotiate with customers than the council ordinance now provides. In their view, the currently permitted contract terms are too restrictive for DWP to meet competition as it may develop. Under a number of competitive scenarios, DWP would lose revenue from business customers, which would put strong pressure on the mayor and council to either reduce DWP payments to the city or raise residential electric rates or do both.

DWP Financing of Other City Operations

Beyond transferring 5 percent of gross operating revenue to the city general fund, DWP subsidizes street lighting and the energy needs of other city departments. It also subsidizes a variety of city services and operations. These costs, which translate to higher electricity rates, may not be fully sustainable under competition.

City use of DWP-owned real estate at below-market price represents one important category of subsidy. Like other large utilities, DWP owns hundreds of valuable real estate parcels, both developed and undeveloped, which it has purchased with its own revenues. Some of the land and buildings owned by DWP are no longer needed for the department's business. An investor-owned utility would lease, sell, or otherwise dispose of such surplus real estate and invest the proceeds in more directly productive uses. At present, however, DWP must offer its surplus properties for other city uses before they can be sold or leased. Either under formal resolution or informal guidance from the council, many such parcels are claimed by other city departments at significantly less than market rates—sometimes for leases of $1 per year. One such example is DWP's lease of some 62 acres in Chatsworth at $1 per year to the Los Angeles Police Department for use as a firing range. In total, nearly half of the 39 DWP properties listed as surplus in May 1999 have been requested for other city uses (DWP, 1999b).[25] Such transfers save money that would otherwise come from city taxes, but they also represent millions of dollars each year in added costs to DWP and its ratepayers.

DWP's large cash flow also makes it an easy resource for elected officials to tap for funds for budget balancing and other purposes. In the fiscally constrained environment since passage of Proposition 13 by California voters, the department has been asked on a number of occasions to make "supplemental transfers" to the city from sales of assets such as fuel oil or land (McCarley, 1996). Other DWP contributions range from paying for colorful promotional street banners to subsidizing city-sponsored dinners and other events. Although these subsidies benefit the city, they also constitute an added (and hidden) tax on ratepayers and an added burden on DWP in a competitive electricity market.

Overall Problems of Governance for DWP Competitiveness

Each problem described above stems from or is exacerbated by DWP's complex structure of governance. Together, problems can interact to add cost and extend the time required to make decisions or carry out normal business processes. For example, hiring a new marketing manager to one of DWP's 16 exempt positions requires a council resolution and then mayor and council approval of the person selected. If the position is not exempt, the individual must be hired within an established civil service job category and pass a standard civil service examination. The process can easily take several months, which makes it very difficult for DWP to recruit from outside. As a consequence, DWP's marketers are mostly longtime engineering staff with little or no marketing training or experience in competitive industries.

As another illustration, negotiating a new customer contract must stay within the limits set by council ordinance. Drafting it may take a good deal longer than it does for other utilities, because the DWP does not control its own legal staff. Changes from the standard contract format may need approval from the commission, the council, and the mayor (under ED39), which can take weeks or months. While such delays might not be a major concern to a monopoly, they can make the difference between winning and losing a customer in a competitive market.

Perhaps even more important, the general manager and other DWP managers spend most of their working time negotiating approvals within the current divided governance system. As David Freeman describes his experience, "It's hard to understand how inhibiting to entrepreneurship and enterprise this governance system is. It takes forever to get permission to do almost anything." (*Metro*, 1998.) Unlike for-profit firms, where managers increasingly look outward to customers and competitors, DWP management primarily focuses inward to the commission, the staffs of the mayor and council, and other city officials.

Chapter 4
Other Governance Models for Municipal Utilities

While DWP and many municipal utilities operate as city departments, others have different organizational and governance structures. This chapter describes and contrasts five such models:

- Municipal utility reporting to city council (e.g., Austin, Texas; Colorado Springs, Colorado).
- Independent city agency (e.g., Jacksonville, Florida; Knoxville, Tennessee).
- City-owned corporation (e.g., Toronto, Ontario; Safford, Arizona).
- Municipal Utility District (e.g., Sacramento Municipal Utility District).
- Joint Powers Agency (e.g., Southern California Public Power Authority).

MUNICIPAL UTILITY REPORTING TO CITY COUNCIL

A number of cities simplify governance by having the municipal utility report directly to the city council. The Colorado Springs City Charter, for example, designates the city council as the board of directors for the utility. The utility executive director then reports directly to the council. Austin, Texas, as well as a number of California cities—including Burbank, Glendale, and Pasadena—have similar governance structures but include council-appointed citizen advisory commissions.

In 1998 Colorado Springs also adopted a new governance framework "suited to today's business reality in which flexibility, quick responsiveness, and clear long-term direction are essential to success." The framework, largely developed by consultant John Carver,[26] seeks to separate the policy functions of the board from the operational responsibilities of the executive director. The board sets policies and communicates them in writing solely to the executive director; it "will never give instructions to persons who report directly or indirectly to the Executive Director." (Colorado Springs, 1998.)

Board policies set out the utility's purpose and ends to be achieved. They also designate what actions of the executive director are unacceptable to the board, in both general ("any practice . . . which is either unlawful [or] imprudent . . . ") and specific ("he or she may not change his or her own compensation or benefits") terms (Colorado Springs, 1998, Policy Numbers EL-1 and EL-4). The executive director may then make all decisions and carry out any activities not expressly prohibited by the board, without seeking further approval.

Direct reporting to the council seems to work well in smaller cities with utilities of relatively modest size. The model does not seem appropriate for a utility as large and complex as the DWP or for a city as diverse and fractious as Los Angeles. However, many of the governance principles adopted by Colorado Springs—particularly the limits set on council involvement in utility operations—are worth consideration here as well.

INDEPENDENT CITY AGENCY

Jacksonville, Florida, and Knoxville, Tennessee, have municipal utilities that operate as city agencies with strong, independent governing boards (Table 4.1). Board members are appointed by the mayor and confirmed by the council for fixed, staggered terms. Unlike in Los Angeles, board members are expected to serve their full terms—in Jacksonville, removal requires a two-thirds council vote; in Knoxville, members can be removed for cause only by a four-fifths vote of the board. These arrangements promote board continuity and independence.

The JEA (formerly Jacksonville Energy Authority) and Knoxville Utility Board (KUB) exercise strong authority under their city charters to govern municipal utilities. The boards can hire and fire the CEO without approval from the mayor or council. The boards set rates after holding public hearings. They delegate to the CEO virtually all customer contract, procurement, real property management, and personnel matters.[27] Senior management in Knoxville and essentially all managers in Jacksonville are exempt from civil service.[28]

These city councils retain only limited authorities over their utilities. In Jacksonville, the council approves the JEA annual budget and must authorize increases in total utility debt, leaving the approval and details of individual debt issues to the JEA Board. Utility payments to the city, currently set at 5.5 mils per kwh sold, are renegotiated every five years. By contrast, the Knoxville City Council approves individual KUB debt issues, but the board approves the budget. Payments "in lieu of taxes" to the city follow Tennessee state law and are based on net plant value and gross operating revenue. In neither city does the council or mayor exercise control over board agendas, board decisions, utility personnel, or operations.[29]

The Knoxville Charter gives the KUB authority to hire its own legal advisor and staff. In Jacksonville, as in Los Angeles, city attorney staff represents the utility. To hire outside counsel, JEA must obtain approval of the city attorney but not the city council.

The governance systems in Jacksonville and Knoxville were designed to distance utility daily operations from city politics, and they appear to work quite well. JEA and KUB are highly regarded both in their cities and by the U.S. public power community. Although JEA and KUB operate with considerable independence, in each case the board, CEO, and other top managers regularly stay in close touch with the mayor and city council. As one executive told us, "We routinely tell the mayor and council what we're planning and how we're doing, even though we're not legally obliged to do so. . . . That's just good politics and good business."

CITY-OWNED CORPORATION

A third governance model involves "corporatization," that is, changing the utility's organizational structure from a city department to a city-owned corporation. The motivation is to improve utility operations and simplify governance, usually in response to or in anticipation of competition. While most electric utility corporatization has

Table 4.1
Governance Comparisons: DWP and Independent City Agencies

Governance Structure	DWP Under New Charter	Jacksonville Energy Authority	Knoxville Utilities Board
Utility structure and size (1998 electricity revenue in millions)	City department ($2,163)	Independent city agency ($777)	Independent city agency ($296)
Governing board	Five-member commission; five-year, staggered terms Members appointed by mayor, confirmed by council Mayor may remove without council approval	Seven-member board; four-year, staggered terms; two-term limit Members appointed by mayor, confirmed by council Mayor may remove with two-thirds council approval	Seven-member board; seven-year, staggered terms; two-term limit Members appointed by mayor from list of five names submitted by board, confirmed by council Removal only for cause by four-fifths board vote
Board authority	Hires and fires CEO with mayor and council approval	Hires and fires CEO Rate setting Individual debt issues Entering new businesses	Hires and fires CEO Rate setting KUB budget approval Entering new businesses
Authority delegated to CEO	Hiring up to 16 exempt positions with mayor's approval, unless council vetoes by two-thirds vote Customer contracts within council guidelines; Procurement <$150K	Hiring 150 exempt positions and other personnel matters Customer contracts Real property sales/leases Procurement	Hiring 30 exempt positions and other personnel matters Customer contracts Real property sales/leases Procurement
Authority retained by council	Approval of rates Job classification and compensation Procurement >$150K Real property sales/leases New debt authorization Capital project approval Entering new businesses Customer contract guidelines Veto of any commission decision by two-thirds vote Outside legal counsel approval	JEA budget approval Overall debt limits JEA payments to city (negotiated every five years) JEA Charter amendments by two-thirds vote with mayor's approval, four-fifths without	Individual debt issue approval
Legal staffing	Provided by city attorney Outside legal counsel must be approved by council and city attorney	Provided by city attorney Outside legal counsel must be approved by city attorney	Board hires legal advisor
Payments to city	5% of operating revenue Ratepayers pay utility tax	5.5 mils per kwh with minimum base of $58 million in 1998 Ratepayers pay utility tax	Payments "in lieu of taxes," based on net plant value and gross operating earnings

occurred outside the United States—in Canada, the United Kingdom, Germany, Australia, and New Zealand, among other countries—it is of growing interest to U.S. municipal utilities as they prepare for competitive electricity markets.[30] Corporatization of the small municipal utility in Safford, Arizona, was highlighted at the 1999 annual meeting of the American Public Power Association (Mecham, 1999).

The recent corporatization of Toronto Hydro, the second-largest municipally owned utility in North America (after DWP), seems particularly relevant to this discussion.[31] Toronto Hydro was restructured under the 1996 Ontario Energy Competition Act, which requires all municipal electric utilities in the province to incorporate by November 2000. At that time, customers will be able to purchase electricity from competitive suppliers and have their bills unbundled to show separate charges for generation, transmission, and distribution.[32] The Toronto Hydro restructuring also amalgamates the City of Toronto's utility operations with those of six adjacent municipalities.[33]

Under the Shareholder Agreement adopted by the Toronto City Council in June 1999, the city transferred all "employees, assets, liabilities, rights, and obligations" of its municipal utility to the Toronto Hydro Corporation, a corporation established under the Ontario Business Corporations Act with the city as the sole shareholder (Toronto, 1999b). The corporation's 11-member board of directors is appointed by the city council for fixed, staggered terms (Table 4.2). Currently, three city council members and eight other citizens serve as directors. The council may remove or replace directors at any time.

As sole shareholder, the council has rights to amend the corporation's bylaws, change the board structure or share structure, and control any change of ownership, dissolution of the corporation, or sale of "all or substantially all" of its assets. The council also retains authority under the Shareholder Agreement to approve new debt issues, annual capital outlays above $170 million, and any service expansion beyond Toronto Hydro's current territory. Except for these reserved powers, the board has full authority to "supervise the management of the business and affairs of the Corporation."

The board delegates to the CEO "the management of the business and affairs of the Corporation," including personnel, customer contracts, procurement, property management, and the hiring of legal staff and advisors.

When incorporation took place in June 1999, the city received $100 million in cash and $34 million in surplus assets from the corporation (Toronto Hydro, 2000). The city also stipulated that of the assets it transferred to the corporation, about 60 percent constituted debt on which the city will receive interest payments of more than $60 million per year. The city also expects the corporation to pay regular dividends corresponding to two-thirds of gross operating earnings from electricity distribution.[34]

While the Toronto Hydro restructuring is too recent to evaluate in terms of operating results, it appears to be moving ahead after surmounting a number of initial obstacles. Many Toronto citizens objected to the amalgamation bill as having been forced on them by a politically conservative provincial legislature. Labor leaders objected to a companion bill as limiting their right to strike and other worker rights during the transition (Ontario, 1997). The amalgamation required harmonization of some 55 collective bargaining agreements from seven separate municipalities covering nearly 5,000 job classifications. Much in the way of implementation remains to be done. And some saw corporatization as merely a stalking horse for privatization of Toronto Hydro.

The Toronto City Council, however, has affirmed its commitment to operating Toronto Hydro as a city-owned utility. The council's Strategic Policies and Priorities Committee emphasizes the benefits of continued public ownership: "As a major player in the competitive industry, Toronto Hydro could be influential in ensuring that energy conservation and environmental responsibility are retained as important issues for consumers." The committee further recommends "that Council leave open the option for Toronto Hydro to develop and invest in the nonregulated, competitive businesses permitted by legislation whenever there is a good business case, risks are reasonable, and returns are satisfactory. . . . However, care must be taken by Council to permit the new board to operate on a commercially prudent basis if it

Table 4.2
Governance Comparisons: DWP, City-Owned Corporation, and
Municipal Utility District

Governance Structure	DWP Under New Charter	Toronto Hydro Corp.	Sacramento Municipal Utility District
Utility structure and size (1998 electricity revenue in millions)	City department ($2,163)	City-owned corporation $1,246 (U.S.$)	Municipal Utility District ($766)
Governing board	Five-member commission Five-year, staggered terms Members appointed by mayor, confirmed by council Mayor may remove without council approval	11-member board of directors 18-month terms for city councilors, three-year staggered terms for others Members may be replaced at any time by council majority vote	Seven-member board, elected by voters for four-year, staggered terms
Board authority	Hires and fires CEO with mayor and council approval	All powers except those reserved to city council as shareholder	All powers as authorized under the California Municipal District Act of 1921
Authority delegated to CEO	Hiring up to 16 exempt positions with mayor's approval, unless council vetoes by two-thirds vote Customer contracts within council guidelines Procurement <$150K	All personnel matters Customer contracts Procurement Real property sales/leases Hiring legal staff "Management of the business"	Most personnel matters Procurement <$100K Day-to-day management as delegated by board
Authority retained by council	Approval of rates Job classification and compensation Procurement >$150K Real property sales/leases New debt authorization Capital project approval Entering new businesses Customer contract guidelines Veto of any commission decision by two-thirds vote Outside legal counsel approval	Bylaw amendments Board structure Share structure or sales Dissolution or sale of "substantially all" assets New debt issues Approval of annual capital outlays >$170 million Service expansion beyond Toronto Ontario Energy Board must approve rates	Board is legislative body of the district
Legal staffing	Provided by city attorney Outside legal counsel must be approved by council and city attorney	Hired by CEO	Hired by board
Payments to city	5% of operating revenue Ratepayers pay utility tax	Two-thirds of operating cash flow of distribution company Interest on city debt Initial transfer of $134 million on incorporation	The Sacramento Municipal Utility District makes no direct payments, but ratepayers pay utility tax to cities

is to enter the competitive market. The pursuit of a nonprofit agenda could result in a nonviable business." (Toronto, 1999a.)

MUNICIPAL UTILITY DISTRICT

Under California's Municipal Utility District (MUD) Act, county voters can establish a separate public agency to provide electricity, water, transportation, or other utility services countywide or within a specified district of the county. If approved by the voters, such a MUD has the same powers as other public agencies, including powers "to sue and be sued, contract, eminent domain, purchase, issue bonds under several authorizing acts, own property and provide utility works and services." (Beck, 1996c.) A MUD is governed by an elected board of directors, with each director representing a specific ward as set out by the county board of supervisors.

As an illustration of MUD governance in California (see Table 4.2), the Sacramento Municipal Utility District (SMUD) Board of Directors has seven members elected for staggered, four-year terms. Directors must be residents of the wards from which they are nominated. However, every voter in the district may vote for all the directors to be elected. SMUD is subject to the Brown Act, so that board meetings are open to the public and must be held at least once a month.

The board appoints a general manager who serves at its pleasure, and it can create or abolish other positions and set salaries as it sees fit. The SMUD Board delegates most personnel decisions to the general manager, so long as they are in accordance with the district's own civil service provisions. No more than 2 percent of appointments can be exempt from civil service. The MUD Act explicitly states that the board may appoint an attorney who serves as the legal advisor to the district.

The SMUD Board generally has broad authority over the district, including setting public tariffs (after a public hearing) and approving customer and supplier contracts. In 1997, the board approved an economic development discount for Intel Corp. in Folsom, California, whereby Intel's base electricity rate would drop by 25 percent if the company added another 600 jobs in the next two years. SMUD offers similar discounts to other companies. For procurement, awards over $50,000 must be offered to the lowest responsible bidder. The general manager may determine the lowest responsible bidder for contracts of less than $100,000.

SMUD has full authority to incur indebtedness and issue general obligation (GO) or revenue bonds. However, voter approval by a two-thirds margin is generally required for new GO bonds, so that municipal utilities rely on bonds backed by their own revenues. The MUD Act requires a municipal utility district to have eight years of operating experience before it can issue revenue bonds.

In 1997, SMUD became California's first municipal electric utility to offer direct access to some of its commercial and industrial customers. It plans to give all its customers direct access to competitive suppliers by 2002. SMUD's strategy to prepare for competition has been to freeze prices for five years through 2002, keep rates 5 percent lower than competitors', and implement a debt-reduction program (SMUD, 1999).

Although SMUD has much more autonomy than a city department and can respond more quickly to competitive changes, converting DWP into a new MUD in Los Angeles would require political approval at several levels. First, the city council would have to pass a resolution calling for the Los Angeles County Board of Supervisors to hold an election to establish the MUD. The supervisors would then submit the proposal to the Local Agency Formation Commission (LAFCO) for analysis and approval. If approved by LAFCO, the proposal would be placed on the ballot at a county election, while the requisite city charter amendments would be submitted to city voters. If both county and city voters passed these measures, the new MUD could be established. For the new entity to be fully functional, however, the California legislature would then have to pass special legislation to permit the MUD to sell revenue bonds prior to its establishing an eight-year operating history. Converting DWP into a MUD thus would require closely coordinated legislation at the city, county, and state levels, as well as approval from city and county voters. Once established with its separately elected board, a MUD would be well insulated from change or control by other local officials.

Joint Powers Agency

Under the California Joint Powers Act, two or more cities, counties, or other public agencies can create a Joint Powers Agency (JPA) to manage electricity generation and transmission facilities or other utility operations. Each participating agency executes a Joint Powers Agreement specifying the JPA's structure, scope, and powers.[35] The JPA is governed by a board of directors whose members represent the participating agencies and are usually appointed by each participant's governing body.

The Joint Powers Act grants broad authorities to a JPA to own property, incur debt and issue revenue bonds, purchase, contract, sue and be sued, provide utility services and set rates for them, and engage in selected other municipal enterprises. It may participate in a member agency's civil service system, although it is not required to do so. One significant restriction is that a JPA cannot issue revenue bonds to acquire or construct electric or water distribution facilities.

One JPA, the Southern California Public Power Authority (SCPPA), comprises DWP, nine other municipal utilities, and the Imperial Irrigation District. It was formed in 1980 to finance the acquisition of generation and transmission facilities for its members. The 11 SCPPA directors are the general managers of its member utilities;[36] each utility gets one vote. However, on issues concerning specific projects, each utility's vote is weighted according to its financial contribution to the project. This means that a majority stakeholder in a project can effectively dictate SCPPA policies and actions for that project.

SCPPA operates on an annual budget of less than $1 million with a staff of three full-time and 10 contract employees. It is a financing rather than an operating organization, unlike its counterpart, the Northern California Power Authority (NCPA), which has 170 employees, operates power plants, and runs power pools.

A JPA has potential advantages of flexibility and, through its appointed board, some independence from local politics. However, the loss of direct control can make local elected officials less than enthusiastic about transferring assets and authorities to a JPA. The restriction against using revenue bonds to acquire distribution facilities also poses a major problem for a utility that intends to offer retail as well as wholesale services. Although some approaches have been suggested to finesse the distribution facilities issue,[37] restructuring DWP into a JPA might well require new California legislation to amend the Joint Powers Act.

Chapter 5
Governance Options for DWP

DWP's existing governance system appears overly complex, cumbersome, and bureaucratic—particularly when compared with other municipal utility structures described in Chapter 4. In the short run, some procedural changes could improve decisionmaking and oversight within the department's current structure. But in light of the city's need to decide whether or not to open the Los Angeles electricity market to competition, serious consideration of more streamlined governance structures for a competitive, municipally owned utility seems in order.

Our review of other municipal utility experience, as well as new governance models discussed in our interviews with stakeholders, suggests two principal alternatives to the status quo:

- Create a city-owned corporation to provide utility services.
- Create a more independent city agency governed by a strong board or commission.

Because of the need for state legislation and for other reasons outlined in the previous section, the options of establishing a new MUD or a JPA appear more difficult to achieve. If passed by county voters, however, a MUD would also be the structure most resistant to change by or influence from local elected officials.

Creating either a city-owned corporation or a more independent city agency with a strong governing board would be controversial and would require amending the city charter significantly. Consequently, we also explore a third, but probably less effective, option:

- Modify the existing structure to improve DWP governance.

These three options are discussed below, and their basic governance functions are compared with the status quo in Table 5.1.

Table 5.1
Governance Options for DWP

Structure and Governance	DWP Status Quo (New Charter)	City-Owned Corporation	Strong Board or Commission	Modified Status Quo
Structure	City department	Corporation[a]	Independent Agency[a]	City department
Governing board term of office, selection, and removal	Five-member commission; five-year, staggered terms Mayor appoints, council confirms Mayor may remove without council approval	Board of directors, seven to nine members with staggered terms[a] Mayor appoints, council confirms Mayor may remove with two-thirds council approval[a]	Seven to nine member commission with staggered terms[a] Mayor appoints, council confirms Mayor may remove with two-thirds council approval[a]	Five-member commission; five-year, staggered terms Mayor appoints, council confirms Mayor may remove with council majority approval[a]
General manager or CEO	Board appoints with mayor and council approval Board may remove with mayor's approval unless vetoed by two-thirds council vote	Board appoints and may remove[a]	Board appoints and may remove[a]	Same as status quo
Other employee status	Civil service except for 16 exempt positions, requiring mayor's approval unless vetoed by two-thirds council vote	Corporation operates separate personnel system[a]	Agency operates separate personnel system under general oversight of council[a]	Same as status quo, but with more delegation to DWP management and commission
Legal staffing	Provided by city attorney; council also approves use of outside counsel	Board appoints legal advisor[a]	Board appoints legal advisor[a]	Board appoints legal advisor with city attorney and council approval[a]
Authorities retained by city council	Approve rates Authorize new debt Approve joint capital projects Job classification and compensation Approve procurement >$150K Set customer contract guidelines Approve property sales/leases Can veto any board decision (Prop. 5)	Approve rates Authorize new debt Represent city as sole shareholder, but no Prop. 5 veto power[a]	Approve rates Authorize new debt General policy oversight of board and agency, but no Prop. 5 veto power[a]	Same as status quo, but with more forbearance and delegation to DWP management and commission
Other limits on board authority	CAO approves agenda items (ED39) Open board meetings under Brown Act	Open board meetings under Brown Act	Open board meetings under Brown Act	Open board meetings under Brown Act
Payments to city	5% of operating revenue	5% of operating revenue	5% of operating revenue	5% of operating revenue

[a] Requires city charter amendment.

OPTION 1: A CITY-OWNED CORPORATION TO PROVIDE UTILITY SERVICES

Forming a city-owned corporation, with a strong board of directors removed from day-to-day city politics, would give DWP much greater flexibility, encourage operating efficiencies, and enable the utility to respond more quickly to the marketplace. The recent incorporation of Toronto's municipal utility shows that this is a realistic alternative if desired by Los Angeles elected officials and voters.

Under this option, voters would be asked to approve city charter amendments that would transfer DWP's assets and operations to a newly formed California nonprofit corporation,[38] governed by a board of directors with the city as sole shareholder. For clarity, we describe here a specific governance structure, but the corporate form is very flexible and would permit a variety of other implementations.

The corporation's board of directors would have authority to hire and fire a general manager or CEO—to whom it would delegate day-to-day decisions—as well as a chief legal advisor. The board would be responsible for overseeing all of the utility's operations except for those expressly reserved to the mayor, council, or other city officials. The council would retain its power to approve rates, authorize new debt, and generally represent the city as sole shareholder. However, it would not have Prop. 5 veto power over board decisions.

Political accountability of the utility to the city's elected officials would primarily reside in the power of the mayor and council to appoint and remove directors. The most straightforward approach is for the mayor to appoint and the council to confirm all directors, although dividing appointments between mayor and council or other variations would be feasible.[39] Terms would be staggered, and removal of a director by the mayor would require a supermajority vote of the council. We would recommend expanding the board to at least seven and up to nine members to be representative of the city and encompass the business, financial, and other skills required to oversee the utility's management and operations. However, we do not recommend earmarking board seats for specific qualifications or constituencies.

Unless specifically restricted by charter or ordinance, a city-owned corporation would have more flexibility than a city agency to attract

managers and skilled professionals, offer financial incentives for performance, negotiate with customers and suppliers, and operate in a
businesslike fashion. It would use generally accepted accounting principles (GAAP) and file financial reports, as other corporate utilities do.
Corporatization would enable Los Angeles's municipal utility to compete most effectively head-to-head with investor-owned utilities and
other private corporations while retaining the benefits of public ownership (Poole, 1998).

This option would require voter approval of several amendments
to the city charter as well as the necessary implementing ordinances.
The charter amendments could include a provision to maintain utility
payments to the city's general fund at 5 percent of operating revenue,
unless limited by bond covenants or other charter provisions. And to
dampen concern that corporatization might inexorably lead to privatization, the charter could require voter approval for any change in
the city's 100 percent ownership of the utility.

OPTION 2: AN INDEPENDENT CITY AGENCY WITH A STRONG GOVERNING BOARD

DWP could operate more flexibly and efficiently as a city agency if
it had a single, strong governing board or commission that was more
insulated from day-to-day politics but still responsive to public concerns. This was basically the model for the Board of Water and Power
Commissioners in the 1925 City Charter. Although the strong commission model no longer exists in Los Angeles (as discussed in Chapter
2), it works well for the municipal utilities in Jacksonville and
Knoxville.

Governance would be similar to that described above for a city-
owned corporation. We would recommend expanding the current
commission to seven to nine members who would be appointed by the
mayor and confirmed by the council, or appointments could be divided
between mayor and council.[40] To provide board continuity and independence, members would serve staggered terms and have some protection against arbitrary removal. In Jacksonville, for example, the
mayor's removal of a board member requires council approval by a

two-thirds vote. As in Option 1, the charter would not specify qualifications for individual commissioners, but the commission as a whole should be broadly representative of the city and have the requisite experience and skills to oversee the utility.[41]

The commission would have full authority to appoint and remove a general manager. Under broad general policies set out by the mayor and council, the commission would be responsible for specific policy for and oversight of all aspects of the utility's normal operations. The commission would delegate day-to-day management decisions to the general manager, subject to board oversight.

The council would retain its authority to approve rates, authorize new debt, and provide general policy guidance to the commission, but it would not have Prop. 5 veto power over commission decisions. On personnel matters, we recommend giving the commission as much flexibility and autonomy as is practically possible.[42]

This option would also require voter approval of several amendments to the city charter as well as the necessary implementing ordinances. The charter amendments could again include a provision to maintain utility payments to the city's general fund at 5 percent of operating revenue, unless limited by bond covenants or other charter provisions.

OPTION 3: MODIFICATIONS OF THE EXISTING STRUCTURE TO IMPROVE DWP GOVERNANCE

Although the currently divided authorities among mayor, council, and commission are cumbersome and often conflicting, governance and decisonmaking could be improved somewhat within DWP's existing structure. The goals would be to focus governance on policy issues, limit political involvement in routine business matters, and streamline approval processes. The changes listed below would not advance these goals as effectively as restructuring under the first two options. However, with two exceptions as noted below, they would not require new charter amendments or other major structural changes.

Expect DWP Commissioners to Serve Full Five-Year Terms

In a more competitive environment, DWP needs knowledgeable and experienced commissioners who can make decisions primarily based on their independent judgment. The commission must remain politically responsible to the mayor and council, but it should be insulated from undue political influences on normal DWP business matters.

Under both the old and new charters, commissioners are appointed to five-year, staggered terms by the mayor and are confirmed by the council. It is a fiction however, that they serve out their full terms—unless the current mayor wants them to. The mayor can remove commissioners at any time under the new charter. And although not required to do so under the charter, commissioners by custom offer their resignations after a mayoral election so that a new mayor can appoint his or her own commission.

To maintain independence and continuity, DWP commissioners could be expected to serve out their terms unless there is cause for their removal. They would not resign when a new mayor is elected.[43] The appropriate model is the Ethics Commission, whose members can be removed by the mayor only with council approval or for cause by a two-thirds vote of the council.[44]

This change would require a charter amendment.

Enable DWP to Hire a Legal Advisor and Staff

An enterprise as large and complex as DWP needs a chief legal advisor who is directly responsible to the board and general manager. DWP also needs attorneys with specialized knowledge of water, power, and environmental law. The current system, in which the city attorney's office provides legal services to the department, handles many routine matters satisfactorily but does not provide the rapid response and full range of expertise that DWP requires.

The Board of Water and Power Commissioners could be authorized to appoint a chief legal advisor who will report to the general manager and the board. If the current departmental structure is retained, having the city attorney and/or the council approve this appointment may be necessary and appropriate. The legal advisor should be able to hire a small staff and retain outside counsel using the department's funds.

This change would require a charter amendment.

Eliminate Formal CAO Review of Commission Agenda Items

While citywide coordination of the departments is an important function of the mayor's office, the current use of ED39 to control commission agendas seems heavy-handed. In DWP's case, requiring formal CAO review and approval of individual agenda items under ED39 results, for the most part, in unnecessary paperwork and delay.

Coordination can surely be handled less bureaucratically. The mayor's staff assigned to monitor each commission can readily obtain the agenda before a commission meeting. If the mayor's office has a problem with an agenda item, staff can request that the item be continued or removed from the agenda. Regular, informal consultation with the mayor's office rather than formal ED39 review would speed up and, we believe, improve the decisionmaking processes of the Water and Power Commission.

This change can be made by the mayor. It does not require council or voter approval.

Eliminate Council Oversight of DWP Routine Business Matters

Even though the existing structure mandates council oversight of DWP operations, a more competitive environment will demand more delegation of authority and some forbearance. Specific steps for the council to consider include giving DWP management more flexibility to

- make larger procurement awards (up to $500,000 or $1 million);
- negotiate long-term customer contracts within a broader range;
- sell or transfer surplus real property;
- expedite hiring of technical and marketing employees; and
- exercise greater autonomy in other personnel matters.

Most important, the council can forbear using its review and veto powers under Prop. 5 for oversight of DWP contracting and other business decisions. Whatever the outcome of council reconsideration, Prop. 5

has a chilling effect on DWP management and commission decisions. The threat of Prop. 5 veto fosters bureaucratic delay, takes up scarce management time, encourages more paperwork to justify decisions and adds uncertainty to normal business dealings. The impacts appear to be significant and inimical to doing business in the private world.

These changes can be implemented by the council.[45]

DISCUSSION: RATIONALE FOR AND OBJECTIONS TO RESTRUCTURING

All three options discussed above maintain the primary public benefits of a municipal utility:

- Local ownership.
- Local rate setting authority.
- Tax-exempt financing and preferences in purchasing federal power.
- Exemption from most income, property, and business taxes.
- Sensitivity to local economic development, the environment, and other social goals.
- Commitment to make direct transfers to the city's general fund.

Each option also keeps the utility's governing board accountable to city elected officials.

The corporatization and independent city agency options would go furthest in helping DWP become more efficient, businesslike, and responsive to changing market conditions. However, these options require substantial changes from the status quo to invest primary governance responsibility in a single board. Restructuring would give the new board considerably more authority to oversee the utility and would deliberately distance the board and utility from day-to-day oversight by the mayor and council.[46]

A variety of objections to such restructuring were raised during our interviews. Some people believe that, after downsizing and with its debt-reduction programs, DWP operates well enough today. Despite divided governance and relatively long decision processes, they argue, DWP's advantages of low-cost debt and exemption from income and property taxes will enable it to compete effectively for the foreseeable future. Because the system works adequately (if creakily) now, it does not need to be fixed.

Others contend that many of the governance problems DWP perceives are exacerbated, if not caused, by poor staff work and inadequate communication between the department and the mayor's office, CAO, CLA, and council. Some in city government believe that DWP does not "work the system" nearly as well as it could and should. Moreover, they say, DWP does not have the internal competence in finance, personnel, and other areas to function as an independent entity.

A third recurring theme is that the current governance system maintains the checks and balances needed to ensure that DWP serves overall city objectives. The department's mission is to supply water and power services reliably and at low cost, but it also provides other public benefits, such as supporting community activities, requiring contractors to comply with the city's Living Wage Ordinance, and giving other city departments first crack at surplus properties. At the request of elected city officials, DWP also helps fund other public projects. Close oversight by the mayor's office and council thus is necessary to see that such broad, citywide objectives are achieved. A variation on this theme states that public agencies must be held to higher ethical standards than are private enterprises, which again requires close accountability to elected officials and their staffs.

Finally, some argue that giving DWP and its board more autonomy would set a poor precedent and encourage other departments and commissions to seek independence from the mayor and council. If the Water and Power Commission were freed from ED39 and Prop. 5 review, what would prevent the Airport and Harbor Commissions, and indeed all city commissions, from receiving the same exemptions? According to this argument, changing DWP's governance structure would inevitably lead to agency balkanization and the loss of citywide coordination by the mayor and council.

Some of these points can be discussed in concrete terms—e.g., what additional internal financial staff capacity would DWP need to operate as a corporation? But others raise a more fundamental issue about the rationale for public utility ownership in Los Angeles. Should

DWP be run as a public business, emphasizing reliable service, low consumer rates, and cash transfers to the city, while still providing local economic development and environmental leadership, or should it also serve broader social and political agendas? The current governance system tries to meld these two philosophies, and it has successfully done so with DWP as a monopoly provider. In a competitive environment, however, the two goals will increasingly clash, and the city might not be able to satisfy both.

Chapter 6
What Comes Next?

Electricity deregulation has suddenly become a political hot potato. The hot summer of 2000 brought unexpectedly higher demand for power in California, with blackouts in the Bay Area, sharp spikes in the wholesale price of electricity, and a doubling of retail prices for customers of San Diego Gas & Electric, the first investor-owned utility to be fully deregulated. Retail prices for customers of other IOUs also may rise as the transitional rate ceilings under the state deregulation plan phase out by December 31, 2001. As a consequence, customers and government officials fought successfully this summer to cap wholesale spot prices, and some are demanding that deregulation be rescinded or at least substantially revised (Brooks, 2000; Smith, 2000; Vogel, 2000).

Throughout the crisis, DWP not only kept rates stable and continued to reduce debt, but it has earned substantial profits by selling electricity it generates to the California Independent System Operator. Whereas the IOUs sold off their generating facilities under the state deregulation plan, DWP still maintains substantial reserves. In August 2000, DWP persuaded the council to authorize selling its share in the Mohave coal-fired power plant in favor of a plan to upgrade cleaner generating plants within the L.A. basin. A recent opinion piece in the *Los Angeles Times* called DWP "The Unexpected Hero in a Deregulated Electricity Market" and recommended that the city "say 'no' to deregulation" (Erie and Phillips, 2000).

Is it necessary or desirable to deregulate electricity prices and permit direct access competition in Los Angeles? In the short run, the answer is clearly no. But in the longer run, if and when the wild price fluctuations observed this summer settle down and orderly electricity markets again become the norm, the question will arise again. We believe this is likely to occur in the 2002–2004 time frame as new generation capacity comes into service in California, and the IOUs and the California ISO gain more experience in stabilizing electricity markets.[47] While city, state, and federal governments stand ready to act to protect consumers from large, short-term price increases, we do not expect them to reverse the underlying trend toward encouraging more competition among electricity suppliers.[48]

Whether or not the Los Angeles electricity market is opened to competition, DWP's governance needs simplification and streamlining. We believe that the restructuring options of corporatization or governance by a strong commission deserve serious attention. Under its current governance structure—even with the modifications recommended in Option 3—DWP would find itself severely constrained in meeting the competition from more agile private firms that we expect to emerge around 2002–2003. In principle, a city-owned corporation (Option 1) could have more operational flexibility than an independent city agency (Option 2), but either form would support faster decision-making and greater responsiveness than does the present structure. Establishing a single governing board, with clear authority and considerable independence from day-to-day political influences, seems a prerequisite for success in a more competitive marketplace.

The issues of DWP governance and possible restructuring are necessarily linked to the council's consideration of whether or when to open the Los Angeles electricity market. Even if the council's decision on direct access is not made until 2002 or later, discussion and debate should begin soon. The public needs to become aware of issues that, while often technical and complex, will directly affect them as taxpayers and ratepayers. The effects of competition on DWP, its employees, its customers, and the city as a whole need to be more fully explored. Possible charter amendments need to be vetted by the council before they can be put before the voters. These issues also are likely to arise in city council races and in the mayoral election of 2001.

Whatever the council's decision on direct access competition, DWP must improve its decisionmaking pace and processes. It must run faster in the future to stay competitive. Strengthening its governance structure seems essential to ensuring reliable electricity supplies, low rates, and adequate payments to the city, as well as to maintaining Los Angeles's leadership among the nation's municipal utilities.

Appendix: A Brief History of DWP[49]

In December 1902, the voters of Los Angeles created a Water Department by amending the city charter. The amendment placed control of the department in the hands of an appointed, five-member commission. The commissioners were to be appointed by the mayor and confirmed by the council. They were to serve staggered, four-year terms and no more than three of them were permitted to be members of the same political party. No one could be appointed a commissioner who had not resided in the city for five years.

Burton Hunter (1933, pp. 105–106) described the governance of the Water Department in a book published in 1933:

> This board elected its own president, who served for one year at a salary of $3000 and was the executive officer of the water department. The superintendent of waterworks, the water overseer, the secretary and all employees of the department were appointed by the board, which had the power to determine the number of its employees, fix their hours of work and rates of pay and require bonds from any or all of them.
>
> All moneys received from the sale or use of water were placed in the water revenue fund and such fund was under the complete control of the board, except that the council might apportion by ordinance such amount as was required to meet principal and interest payments on outstanding waterworks bonds. Also while water rates were fixed by the commission, the approval of the council was required.[50]
>
> Three members constituted a quorum; but a vote of three was required on any action involving the making of a contract, auditing a bill, expending money or incurring debt. The city auditor handled demands on the water revenue fund in the same manner as those on the school and library funds, demands re-

quiring the signature of the president, or in his absence two members, and the secretary. Competitive bidding on purchases over $500 with award to the lowest responsible bidder was mandatory. A monthly report and an annual report to the council were required.

The chief engineer of the Water Department began to look for an additional source of water in 1904. In 1905, the city spent $1.5 million to purchase water rights in the distant Owens Valley, looking forward to building an aqueduct to the city. In 1907, the city approved a $23 million bond issue to pay for aqueduct construction. It was not until after the city had approved the 1905 and 1907 bond issues that leaders began to consider the electrical power that would be generated as a by-product of the Owens Valley Aqueduct. In February 1909, the voters passed charter amendments to begin to deal with disposal of the hydroelectric power the city would soon possess. The amendments empowered the city "[t]o provide for supplying the city with . . . electricity . . . or with other means of heat, illumination or power; and to acquire or construct and to lease or operate . . . plants and equipments for the production or transmission of . . . electricity, heat, refrigeration or power, in any of their forms, by pipes, wires or other means; and to incur a bonded indebtedness for any of such purposes." (*Charter of the City of L.A.*, 1909, Section 7.)

In early fall 1909, the council appropriated $10,000 to finance preliminary work and established a Bureau of L.A. Aqueduct Power. E. F. Scattergood was made the chief electrical engineer, and a board of consulting engineers was appointed. In September 1909, the Board of Public Works recommended to the council a $3.5 million power bond issue. The council set April 10, 1910, as the date for the power bond election, and the voters gave the bond issue the necessary two-thirds approval. On March 6, 1911, the public voted for municipal distribution of electricity, rather than sale of the city's power to private power companies, by approving a straw ballot measure.

At the same election, the public voted to create a Department of Public Service with both a Water Bureau and a Power Bureau. The Power Bureau had its own chief engineer and general manager, reporting directly to the Board of Public Service Commissioners. The Power Bureau had its own divisions for personnel, public relations, building, maintenance, land and right-of-way acquisitions, and library. Each bureau was a "self-contained administrative unit." (Van Valen, 1964, p. 63.)

The Department of Public Service was headed by a board of commissioners very similar to the Water Commissioners that had preceded it. The only changes were that board members were to be appointed "without regard to their political opinions but with regard to their fitness." (Hunter, 1933, p. 131.) A "newly created power revenue fund" was added to the charter. "The provisions relative to the water revenue fund remained the same as originally written in 1903. The power revenue fund was created along similar lines, except that portions of it might be used for extending the business pertaining to electric power. . . . [In addition, t]erms were provided under which contracts might be entered into to supply municipalities with water and electric energy, as well as the sale of surplus to consumers outside the city." (Hunter, 1933, pp. 131–132.)

The 1911 charter amendments essentially provided for treating the city's electric business in much the same way as its water business. With respect to city-owned utilities, then, the charter remained very similar to the way it had looked in 1903. There was a substantial difference, however, between the operation of the Water and Power Bureaus in practice. The Water Bureau generally had no difficulty in obtaining public approval of the general obligation bond issues it needed to extend the city's ownership of necessary water resources. However, the Power Bureau faced constant challenges at the hands of the private power companies doing business in the City of Los Angeles. As a matter of fact, the April 1913 bond issue with which the Power Bureau sought to supplement the insufficient bond issue of 1910 was defeated after a vigorous campaign by the private power companies—Los Angeles Gas & Electric and Southern California Edison. The bond issue succeeded when it was resubmitted in May 1914, but only because of support from the Los Angeles Chamber of Commerce. Such support would be a precondition for bond issue passage until 1947, when the

Power Bureau acquired revenue bonding authority from a charter amendment.

However, the structure of the Power Bureau changed in more ways than simply in revenue bonding authority between 1911 and 1947. The most substantial change to the Power Bureau and the Department of Public Service to which the bureau belonged to came with the 1925 Charter. The new charter the city enacted in 1925 saw substantial changes in terms of governance of the city's utility business. The 1925 Charter put a halt to the leadership of the department by the salaried president of the board of commissioners. The new charter made the board more like a board of directors and the general manager more like a CEO. The Public Service Department was renamed and became the Water and Power Department.

> The board of water and power commissioners, five appointed citizens with five-year term and fee fixed by charter, succeeded the board of water commissioners and the board of public service commissioners. It holds for the city and controls all water, water rights, electric, the right to develop power, and all lands, buildings and structures in connection therewith. It regulates the use, sale and distribution of water and power, fixing rates subject to ordinance approval; controls all water and power bond funds, the water revenue fund and the power revenue fund; makes its own budget; determines the number of its own employees; fixes salaries; and appoints a general manager, secretary and chief accountant. An annual report is made to the mayor and council. . . .

> The board may divide the department into two bureaus, the bureau of waterworks and supply and the bureau of power and light, and appoint a general manager, chief engineer, for each bureau. "In case such division is not made, the general manager of the department shall be the chief engineer of the department and shall have recognized ability and broad experience in hydraulic and electrical engineering and the economics of water and electrical utilities. . . ."

> The general manager, chief engineer of waterworks, chief electrical engineer, auditor and cashier are the only positions in the

department exempted from civil service by the charter. The general manager, or general managers, makes all appointments, except those mentioned above as made by the board. . . .

Specific provisions are included in the charter relative to the sale of water and power to other municipalities, firms or persons outside the city. The board may sue and be sued, and require the services of the city attorney free of charge. . . .

All revenue of the department is placed in either the water revenue fund or the power revenue fund. Expenditures from such funds can be made only in connection with the respective—water or power—purposes for the operation and extension of works, extension of business, a pension system for employees and the payment of indebtedness. Any surplus may be refunded to the city but only upon approval of the board of water and power commissioners, which is also empowered to negotiate with ordinance approval for emergency loans payable from revenue and not to exceed one-third of the gross operating revenue for the prior fiscal year. (Hunter, 1933, pp. 227–229.)

Of course, the 1925 Charter has been amended extensively since 1925. Most of the amendments enacted from 1925 until the 1960s have eased the business of the DWP. For example, charter amendments in 1935 and 1940 allowed the Board of Water and Power Commissioners to "make temporary arrangements for the interchange or sale of electric power for not more than four years," "make arrangements for the sale or interchange of electric power in connection with the utilization of power from the Colorado River," and "enter into contracts with the national government pursuant to the Boulder Canyon Project Adjustment Act to permit the delivery of electric energy to the city." (Bollens, 1963, p. 119.)[51]

The DWP's autonomy in financial operations was greatly enhanced in these amendments. For example, in 1929, "[t]he need for emergencies is deleted and the department is allowed to borrow if it determines that the demand for service and the financial condition of the works justify doing so. The procedure, terms, and conditions of borrowing must receive the approval of the council and the mayor." (Bollens, 1963, p. 121.) Amendments in 1933 and 1935 allowed the board to "borrow from the national or state government or any authorized agency created by either of these governments" and "borrow from the national or state government to acquire the electric system of the Los Angeles Gas and Electric Corporation." Finally, a 1947 amendment separated the indebtedness of the department "from the general city debt. The department can initiate short-term borrowing whenever it is in the public interest." (Bollens, 1963, p. 121.)

Other amendments during this period gave the department other kinds of useful authority. For instance, a 1937 amendment allowed the department to "establish and maintain a general system of retirement, disability, and death benefits." (Bollens, 1963, p. 120.) Subsequent amendments in 1947, 1951, and 1957 served to "change the procedure and amounts of benefits to be received under the retirement system of the department." (Bollens, 1963, p. 120.) Also, a 1963 amendment allowed DWP "to provide hospital, medical, and surgical benefits to its active and retired employees and their dependents." (Bollens, 1963, p. 121.) These benefits made working for the DWP the most attractive employment opportunity the city could offer and ensured that the department could draw the best civil servants. Another helpful amendment was the 1937 charter amendment that allowed DWP to "extend and promote the electric business of the department through conducting and holding annual expositions." (Bollens, 1963, p. 121.)

Only one amendment from 1925 to 1963 had the effect of weakening the autonomy of DWP. In 1941, the voters passed an amendment indicating that "[a]n industrial and administrative survey is to be undertaken at least every ten years or sooner at the discretion of the mayor. This survey ascertains if the department is being operated most efficiently and economically." (Bollens, 1963, p. 121.) The degree to which the autonomy of the DWP was sought can be indexed by a "wish list" of amendments that the department requested from the city in a September 10, 1940, communication. The DWP sought "[d]ivorcement of the Department of Water and Power, to a substantial degree, from the city government, thereby establishing such Department as virtually an autonomous entity." (Ingram, forthcoming.)[52]

To achieve the autonomy of this substantial municipal divorce, the department recommended a number of changes. Among these were the removal of "appointing power from mayor and approval power from Council"; instead, the charter would provide "for self-perpetuating board with appointments subject to approval of mayor, and in event of his disapproval, subject to approval by board of appointment." Other amendments would have replaced the city controller with a department controller, the city treasurer with a department treasurer, the city's civil service system with a department civil service system, and the city attorney with a general counsel heading a "separate legal division" for the department (Ingram, forthcoming, pp. 2–4).

The amendments suggested that the department, not the council, should control the term and conditions of sale of real and personal property. The council would also lose power to authorize the establishment of the reserve fund, as well as the approval of contracts awarded without advertising for bids. The council would retain its approval of rates, but would approve them by resolution rather than by ordinance to exempt them from the popular referendum. Other city officers would also have lost authority under the department's ideal charter. For instance, the board would "place surety bonds, rather than city controller," the mayor would no longer need to consent to the department's transfer of funds between budget items, the city's Civil Service Board would relinquish approval of payrolls to the DWP's own civil service board, and the department's general counsel (Ingram, forthcoming) would approve contracts instead of the city attorney.

The DWP's financial authority would have been greatly enhanced by the recommended amendments. First, their changes provided "for interest earned by any funds under the control of the department to be credited to such funds. Under existing sections, interest on revenue funds of department is credited to general fund of city." Second, the changes authorized "creation of reserve fund for purpose of conserving and accumulating money which may be expended for general purposes of department. Authority now exists for establishment of reserve funds for special purposes only." Third, the board's discretion to transfer any and all surplus to the city would be replaced by a requirement of "transfer of 5 percent of gross operating revenues for each fiscal year to general fund." (Ingram, forthcoming, p. 5.) Fourth, the department would be authorized to issue 40-year revenue bonds rather than being limited to 12-year ordinary borrowing and 20-year borrowing from the federal government. Finally, the amendments would have allowed the department to create new budget items during the fiscal year and to appropriate in excess of budget amounts when actual revenues exceed estimates. No provision was made for mayor or council involvement in any of these financial powers.

The DWP did not achieve most of these sought-after changes. In fact, the enactment of some of the more routine amendments they wanted in 1940 was also accompanied by a constraint—the requirement for the decennial industrial and administrative survey. In 1938, Mayor Fletcher Bowron was elected through the recall process, and the DWP had supported his recalled predecessor, Frank Shaw. The trend toward amendments freeing the DWP from oversight by elected officials was effectively ended, and the goal of divorcement from municipal control became chimerical. Mayors Norris Poulson, Sam Yorty, and Tom Bradley—Bowron's three successors in office—would not support DWP's quest for autonomy as earlier mayors had. Charter amendments from the 1960s onward reduced the DWP's autonomy, taking away its independent salary-setting authority and giving the council the ability to overturn every commission decision.

The main charter amendments that have reduced the autonomy of DWP from elected officials are as follows: In 1977, an amendment took away salary-setting authority from DWP and gave it to the council (Section 86). In 1991, the DWP's leasing authority was limited to five years, including option clauses that used to allow for longer cumulative terms (Section 220(6)). Later in 1991, Proposition 5 was passed, allowing the council to review and overturn the decisions of the commissions, including the Board of Water and Power Commissioners (Section 32.3). In 1995, the power to appoint and remove the general manager of DWP was transferred from the board to the mayor and council (Section 79). In 1996, the decennial industrial and administrative survey was changed into an "industrial, economic and administrative survey" to be performed every five years; it was also made clear that this survey must be paid from DWP's own funds (Section 220.3 repealed and Section 396 added).

However, it is not only charter amendments that have reduced the DWP's historic autonomy in recent years. In 1984, Mayor Bradley propounded an executive order—Executive Directive 39—that empowered the CAO to act on behalf of the mayor in controlling the agenda of the Water and Power Commission. Although some current and former city officials believe that ED39 violates other sections of the charter and is thus unenforceable, [53] it is still in effect and used routinely by the CAO on behalf of the mayor.

The other source of the restriction of the DWP's autonomy is custom. It has become customary for the members of the city's commissions, including the Board of Water and Power Commissioners, to resign when a new mayor takes office:

> This procedure had its origin in Mayor [John C.] Porter's demand that all commissioners submit their resignations to him at the time he assumed office.

> Elected on a reform platform, Bowron had promised a complete removal of all Shaw appointees, and therefore demanded the resignation of all commissioners. At the time of Poulson's inauguration the concept of commissioner responsibility had reached the point where most resignations were presented to him without any request on his part.

> This practice has weakened the claim that under the existing system of overlapping board appointments a mayor might wait two years before he had a majority of his own appointees on a board to support his policy proposals. Under the conditions described he usually can accomplish this within a few weeks after assuming office.

> It should be noted that this custom is completely contrary to the intentions of the framers of the charter. They created the board form to provide a continuity of policy and insure independence from a new mayor's control for at least two years. In essence the change has moved the pattern of operation closer

to that of a single headed department, and consequently has strengthened the office of mayor. (Abrahams, 1967.)

The days of a strong Board of Water and Power Commissioners, which would press for charter amendments like those on the 1940 DWP "wish list," are gone. The convergence of these changes—in customs, the implementation of ED39, and charter amendments such as Prop. 5—has subjected the DWP to the control of Los Angeles's elected officials.

Endnotes

1. Figures are for the fiscal year ending June 30, 1999.

2. The origins and history of DWP are discussed further in Appendix A and more fully in Ingram (1994).

3. As of this writing, however, DWP is earning a good profit by selling power from its generating plants to the CalPX.

4. In fact, DWP officials have always been concerned about competitive pressures from Southern California Edison. Under its zonal pricing system, SCE has set lower rates near Los Angeles city limits than in other areas. In Mayor Fletcher Bowron's administration in the late 1930s, SCE played a major behind-the-scenes role in securing a 5 percent, "surplus" transfer from DWP to the city general fund. Thus, DWP has had to be on its competitive guard (Erie, forthcoming).

5. We also did not examine possible restructuring or governance changes resulting from separation of DWP water and power operations or secession of the San Fernando Valley from the City of Los Angeles.

6. See also McCarthy et al. (1998).

7. The general manager is called "chief engineer" in the current charter and traditionally has had an engineering background. Controversy arose in the mid-1990s when William McCarley, who did not have an engineering background, was appointed as DWP general manager. However, the city attorney supported the legality of McCarley's appointment. The new charter removes the engineering qualifications for the general manager.

8. By charter, the commission—not the general manager—heads the DWP, although its powers are circumscribed as described in this chapter. The commission has "the power . . . to make and enforce all necessary and desirable rules and regulations for the exercise of powers and the performance of the duties conferred upon" it by the charter, "subject to the provisions of this Charter and to such ordinances of the City" that do not conflict with the charter (Old Charter, 1997, Section 78).

9. Previously, council approval was required to remove a commissioner.

10. "The board of each Proprietary Department shall appoint the general manager subject to confirmation by the Mayor and Council, and shall remove the general manager subject to confirmation by the Mayor" (New Charter, 1999, Section 604(a)). However, the general manager may appeal his/her removal and be reinstated by a two-thirds vote of the council within ten days of the appeal (New Charter, 1999, Section 508(e)).

11. "As a general rule, *anything* requiring the approval of the city council should *first* be sent to the mayor for a review pursuant to Executive Directive 39" (emphasis in original). (Dickenson, 1996.)

12. Another stated purpose for ED39 is to give the commissions a citywide, independent analysis of departmental proposals. The mayor also wants to avoid possible embarrass-

ment from uncoordinated actions of "independent" city commissions, which ostensibly was the principal reason Mayor Bradley issued ED39 in 1984.

[13] One question raised during our interviews was whether ED39 would survive a legal challenge, because the charter gives the head of the department—in this case the commission—powers "to supervise, control, regulate and manage the department and to make and enforce all necessary and desirable rules and regulations therefor and for the exercise of the powers conferred upon the department by this charter." (Old Charter, 1997, Section 78.) This could be construed as precluding any mayoral requirement for CAO approval of commission agendas.

[14] In 1997, for example, the CLA was instrumental in the council's rejection of a proposed DWP power-marketing partnership with a commercial joint venture (Duke/Louis Dreyfus L.L.C.), which had been strongly recommended by consultants and the DWP general manager.

[15] Except for commission actions otherwise "subject to appeal or review by the Council." (New Charter, 1999, Section 245(d)(8).)

[16] Notably, all four items in which the council changed the commission's decisions involved contract matters.

[17] Most of those we interviewed thought that the changes in Prop. 5 will result in fewer challenges to commission decisions, although a few believed that the council might be inclined to use its veto power frequently, perhaps leading to extended Ping-Pong matches between commission and council on some items.

[18] Also exempted are "unskilled laborers, including drivers," construction workers on public works projects, part-time employees, and grant-funded employees limited to three years maximum (Old Charter, 1997, Section 111).

[19] This section of the charter also provides for up to 150 additional exempt positions for "Management, Professional, Scientific or Expert Services" for the entire city, including DWP.

[20] City procurement regulations have been a burden in certain time-critical situations. For example, during the oil crisis of 1973–1974, DWP had to get city council approval to bypass the normal procurement cycle in order to purchase needed oil supplies in a fast-moving, sellers' market.

[21] As part of implementing the new charter amendments, the council authorized raising the limit from $100,000 to $150,000, and the commission approved the $150,000 limit on May 16, 2000.

[22] Contracts of more than three years' duration also must be approved by the council.

[23] The Los Angeles City Council has consistently used its rate-setting power to subsidize consumer (i.e., voter) rates with higher rates for business. Until 1996, DWP's business rates were still below those of SCE. However, according to data in the DWP and SCE 1998 annual reports, DWP's commercial and industrial rates in 1998 averaged 9.3 cents per kwh, compared with SCE's 8.3 cents per kwh.

[24] These points were also made in our interviews with DWP customers.

[25] See also Muto (1999).

[26] The Carver governance structure is described at http://www.carvergovernance.com/model.htm.

[27] KUB has had its own personnel system only since 1998, and workers who are disciplined by KUB can appeal to the City of Knoxville civil service system.

[28] The Jacksonville City Council has increased the number of JEA exempt positions several times as part of a financial package, negotiated every five years, which has included increased payments to the city.

[29] As energy prices rose during the 1970s and energy issues became more heated, Jacksonville voters amended the city charter to give JEA greater long-term stability from city politics by requiring a supermajority vote of the council to change the JEA Charter or bylaws. Such changes require a two-thirds council vote with the mayor's approval or a four-fifths vote without the mayor's approval.

[30] The Reason Foundation has studied corporatization of foreign utilities and concludes that this model can and should be applied to the Los Angeles proprietary departments (Poole, 1998).

[31] The Toronto Hydro case is described in more detail in Mahnovski (1999).

[32] The Ontario Energy Competition Act of 1996 (Bill 35) split Ontario Hydro, the large provincially owned utility from which Toronto Hydro formerly purchased all its electricity, into four separate companies providing generation, transmission, financing of stranded assets, and an independent market operator, similar to CalPx and ISO. The Act also established the Ontario Energy Board to regulate competition in the electricity sector.

[33] The merger of utility operations was part of the overall amalgamation of the City of Toronto with six adjacent municipalities on January 1, 1998, under the City of Toronto Act of 1996 (Bill 105). The amalgamation, passed by the Ontario Provincial legislature rather than the cities, was intended to streamline local government and reduce costs.

[34] Under Ontario's tax structure, taking payments as interest on debt rather than as dividends is advantageous to the city.

[35] A city participant may also need to amend its charter to be consistent with JPA utility operations.

[36] Some general managers have delegated their board responsibilities to other utility executives.

[37] One possible approach would be to lease the distribution facilities to the JPA, with an option to purchase them later from retained earnings (Beck, 1996b).

[38] A city-owned nonprofit corporation would not be subject to state income or Los Angeles property and business taxes. Alternatively, the Reason Foundation recommends incorporating as a for-profit corporation, which would pay taxes to the state and city, with a commensurate reduction or elimination of direct transfers to the city's general fund (Poole, 1998; Moore, Poole, and Woerner, 1998). Tax issues are complex and require further study before any decision can be reached about incorporating as a nonprofit or for-profit entity.

[39] In Toronto, three city council members serve on an 11-member board of directors. This familiarizes the council members with the technical and operating issues involved in running a complex municipal utility, without giving them majority control of the board. David Freeman's 1998 proposal to the Charter Reform Commissions calls for a seven-member board, two appointed by the mayor, two appointed by the council, and three elected by the voters.

[40] Water and Power Associates, an organization of retired DWP managers, has recommended that the mayor appoint three members and the city council president appoint two

members of a strengthened five-member commission. In Knoxville, the mayor selects a new board appointee from a list of five names submitted by the sitting board.

41 The Jacksonville and Knoxville Charters preclude public officials from serving on their utility boards. Knoxville also does not permit "employees or retirees of current or potential energy suppliers" to serve. Both cities require board members to be city residents and limit their board service to two terms.

42 As noted in Chapter 4, the utility in Knoxville runs its own personnel system, but employees who have been disciplined can appeal to the city civil service system.

43 Nor should newly appointed commissioners be asked by the mayor for a signed letter of resignation when appointed, as reportedly has been the custom in some previous city administrations.

44 Cause is defined as "substantial neglect of duty, gross misconduct in office, inability to discharge the powers and duties of office or violation of this Article." (New Charter, 1999, Section 700(e).)

45 Under this option, we would not at this time recommend a charter amendment to rescind or amend Prop. 5 as it applies to the Water and Power Commission. Instead, we would advise council forbearance and "watchful waiting" under the new charter to see whether or how the new Prop. 5 legislative veto affects DWP operations.

46 However, it should be noted that any restructuring by city charter amendment can also be undone by subsequent amendments. Los Angeles voters have approved several charter amendments over the past 25 years that have weakened the independence of the Water and Power Commission and given more power to the mayor and council. As a consequence, some would argue in favor of restructuring into a MUD that, although more politically difficult to enact, would also be more difficult to unravel.

47 The volatility of wholesale electricity prices this summer was due to a number of supply and demand factors acting together: hot weather, a buoyant economy, increased power demand from computer-intensive businesses, aging and thus less reliable power plants, insufficient transmission capacity, and a poorly functioning wholesale bidding market. Although industry and government are addressing these problems, not all of them can be fixed quickly. Substantial wholesale price fluctuations thus seem probable at least through 2001. However, more than 20 new power plants are under development in California and, assuming reasonable improvements in the ISO wholesale bidding system, should help stabilize the supply-demand balance beginning in 2002. For more details, see CPUC, 2000.

48 Most of the city officials and other stakeholders we spoke with thought that eventual opening of the city's electricity market to competition is inevitable if the IOUs remain subject to competition and deregulation. A few dissented, however, saying that so long as DWP provides reliable power at competitive or near-competitive rates, the city council will not have to open the market.

49 This appendix was prepared by James W. Ingram III from research for his Ph.D. dissertation, "A Virtual Reform Machine: Charter Reform in Los Angeles," University of California, San Diego, forthcoming.

50 Despite this financial control, the Water Commission could not convey, lease, or otherwise dispose of water or water rights without the approval of two-thirds of the voters (1903 Charter, Section 191).

51 Both of the 1940 amendments were part of the 1940 DWP "wish list" discussed below. They were among the less significant requests in terms of enhancing DWP's authority and autonomy.

[52] See also Haynes Papers (1940).

[53] Old Charter Section 51(6) states, "The powers and duties of the City Administrative Officer and the provisions of this section shall not apply to the Departments of Water and Power, Harbor, or Airports." Moreover, Section 52 provides that the DWP is exempt from the CAO's authority to make temporary transfers of personnel to deal with shortages. These are the two most important sections in laying out the CAO's jurisdiction. However, no legal challenge has been mounted against the CAO's use of ED39 to review commission agenda items.

Bibliography

Abrahams, Marvin, "Functioning of Boards and Commissions in the Los Angeles City Government," Ph.D. dissertation, UCLA, Los Angeles, 1967.

Barrington-Wellesley Group, Inc., "A Diagnostic Audit of the Los Angeles Department of Water and Power," May 1994.

Beck, R. W., & Associates, "Comparative Forms of Ownership," report, 1996a.

_____, "Joint Powers Agency Alternative for Municipal Electric Utilities," in "Comparative Forms of Ownership," 1996b.

_____, "Municipal Utility District Alternatives for Municipal Electric Utilities," in "Comparative Forms of Ownership," 1996c.

Bollens, John C., *A Study of the Los Angeles Charter: A Report to the Municipal and County Government Section of Town Hall*, Los Angeles: Town Hall, 1963.

Brooks, Nancy Rivera, "Consumers Seek Repeal of Utility Deregulation," *Los Angeles Times*, July 29, 2000.

California Public Utilities Commission (CPUC), "California's Electricity Options and Challenges," http://www.cpuc.ca.gov/published/report/GOV_REPORT.htm, August 5, 2000.

Carver, John, "Policy Governance Model," accessed at http://www.carvergovernance.com/model.htm, 1999.

Charter of the City of Los Angeles (as amended up to February 1909), Los Angeles: Harry M. Wier & Co., 1909.

Charter of the City of Los Angeles (1997 as amended since 1925 [referred to as "Old Charter" herein]), Los Angeles: Brackett Publishing, 1997.

Charter of the City of Los Angeles (1999, as adopted June 1999 [referred to as "New Charter" herein]), Los Angeles: City Clerk, 1999.

Charter Reform Commission (City of Los Angeles), "Discussion Materials," January 14, 1998a.

_____, "Summary of Remarks by S. David Freeman," March 11, 1998b.

Colorado Springs (City of), Colorado., "Colorado Springs Utility Board Policy Governance," adopted October 21, 1998.

Department of Water and Power (DWP) (Los Angeles), *1998–99 Annual Report*, 1999.

_____, "DWP Action Plan to Meet the Competitive Challenge," November 10, 1997.

_____, "Doing Business with the DWP," http://www.ladwp.com/purchasing/about/pabout.htm#general, 1999a.

_____, "DWP Surplus Properties," report, May 24, 1999b.

_____, "Department of Water and Power Signs 22 Industrial Customers to More than $1 Billion in Long-Term Energy Contracts," press release, October 4, 1999c.

Dickenson, R., "General Guidelines for Mayor and Council Review: Proprietary Departments," City of Los Angeles, Office of the Chief Administrative Officer, January 25, 1996.

Edison International, *1999 Financial and Statistical Report*, Rosemead, Calif., 1999.

The Elected Los Angeles Charter Reform Commission, Committee on Improving the Delivery of City Services, "Report on Proprietary Departments," May 26, 1998a.

_____, "Financial Relationship of Airports, Harbor, and Department of Water and Power to the City of Los Angeles," May 26, 1998b.

_____, Task Force on the City Attorney's Office, "Report," August 25, 1998c.

_____, "Task Force and Staff Report on Contracts, Procurements, and Leases," October 19, 1998d.

Erie, Steven P., "How the Urban West Was Won: The Local, State, and Economic Growth in Los Angeles, 1880–1932," *Urban Affairs Quarterly*, Vol. 27, 1992, pp. 519–554.

_____, *L.A.'s Crown Jewels: Public Enterprise and the Politics of Growth and Globalization in Twentieth Century Los Angeles* (working title), Stanford, Calif.: Stanford University Press, forthcoming.

Erie, Steven P., and Robert V. Phillips, "The Unexpected Hero in a Deregulated Electricity Market," *Los Angeles Times*, September 10, 2000.

Freeman, S. David, remarks to the Los Angeles Elected Charter Reform Commission, May 8, 1998.

Haynes Papers, "Plan of Proposed Charter Amendments to Be Submitted by the Board of Water and Power Commissioners," Los Angeles Charter 1940–41 Amendments (proposed), September 10, 1940.

Hunter, Burton, *The Evolution of Municipal Organization and Administrative Practice in the City of Los Angeles*, Los Angeles: Parker, Stone & Baird, 1933.

Ingram, James W., III, "Building the Municipal State," paper presented at the annual meeting of the American Political Science Association, 1994.

_____, "A Virtual Reform Machine: Charter Reform in Los Angeles," Ph.D. dissertation, University of California, San Diego, forthcoming.

Mahnovski, Sergej, "Toronto Hydro: Transition from Municipal Utility to Private Corporation," RAND, unpublished memorandum, September 1999.

McCarley, William, letter from the DWP general manager to the Board of Water and Power Commissioners, November 18, 1996.

McCarthy, Kevin F., Steven P. Erie, and Robert E. Reichardt with James W. Ingram III, *Meeting the Challenge of Charter Reform*, Santa Monica, Calif.: RAND, MR-961-LABA, 1998.

Mecham, Kenneth D., "Safford's New Model for Competitive Governance," presentation to the American Public Power Association, June 21, 1999.

Metro Investment Report, "LA's DWP Hasn't Had a Leader Like David Freeman Since Mulholland," July 1998.

Metzler, Richard, & Associates, "Decennial Survey of the Department of Water and Power," City of Los Angeles, Board of Water and Power Commissioners, 1990.

Moore, Adrian, Robert W. Poole, Jr., and Jeff Woerner, "Water & Power Can Lead the Way to Change," *Los Angeles Times*, December 16, 1998.

Muto, Sheila, "Utility's Hope for Windfall in Real Estate Gets Tested," *The Wall Street Journal*, November 10, 1999.

Nemec, Richard, "Planning for the Future," *Los Angeles Daily News*, March 28, 1999.

Newton, Jim, "DWP Chief Says Deregulation Could Spell Ruin," *Los Angeles Times*, November 23, 1998.

Ontario Provincial Legislature, Ontario Public Sector Transition Stability Act of 1997 (Bill 136), 1997.

Poole, Robert W., Jr., "Reforming Municipal Enterprises: Turning L.A.'s Proprietary Departments into World-Class Competitors," statement before the Los Angeles Elected Charter Reform Commission, May 6, 1998.

PSC Energy Corp., "LADWP Governance Plan (Draft)," December 17, 1996.

_____, "End of Engagement Report," February 4, 1997.

_____, "LADWP 1997 Strategic Business Plan," February 7, 1997.

Riordan, Richard, letter to the Commissioners of the Elected and Appointed Los Angeles Charter Reform Commissions, May 13, 1998.

Sacramento Municipal Utility District (SMUD), "General Manager's Report and Recommendations on Current Rate Issues," January 21, 1999.

Smith, Rebecca, "California Cuts Price Cap for Electricity, Again," *The Wall Street Journal*, August 2, 2000.

Southern California Edison Company (SCE), *1998 Annual Report*, 1998.

State of California, "Public Utilities: Electrical Restructuring," AB 1890, 1996, accessed at http://www.leginfo.ca.gov/pub/95-96/bill/asm/ab_1851-1900/ab_1890_bill_960924_chaptered.html.

Toronto Hydro Corporation, *1999 Annual Report*, 2000.

Toronto, Ontario (City Council of), "Recommendations of the Strategic Policies and Priorities Committee," June 1, 1999a.

_____, "Incorporation of the Toronto Hydro Corporation," adopted June 9, 10, and 11, 1999b.

Van Valen, Nelson, "Power Politics: The Struggle for Municipal Ownership of Electric Utilities in Los Angeles, 1905–1937," Ph.D. thesis, Claremont, Calif.: Claremont Graduate School, 1964.

Vogel, Nancy, "PUC Caps Electric Bills of Some San Diego Users," *Los Angeles Times*, August 22, 2000.

Water & Power Associates, "Recommendations for DWP Governance," report, 1998.

World Bank, *Bureaucrats in Business: The Economics and Politics of Government Ownership*, Oxford, U.K.: Oxford University Press, 1995.